PERGAMON INTERNATIONAL LIBRARY
of Science, Technology, Engineering and Social Studies
*The 1000-volume original paperback library in aid of education,
industrial training and the enjoyment of leisure*
Publisher: Robert Maxwell, M.C.

ATMOSPHERIC POLLUTION

Its History, Origins and Prevention

FOURTH EDITION

THE PERGAMON TEXTBOOK
INSPECTION COPY SERVICE

An inspection copy of any book published in the Pergamon International Library will gladly be sent to academic staff without obligation for their consideration for course adoption or recommendation. Copies may be retained for a period of 60 days from receipt and returned if not suitable. When a particular title is adopted or recommended for adoption for class use and the recommendation results in a sale of 12 or more copies, the inspection copy may be retained with our compliments. The Publishers will be pleased to receive suggestions for revised editions and new titles to be published in this important International Library.

Other Related Pergamon Titles of Interest

BOOTH
Industrial Gases

COOPER et al
Particles in the Air (A Guide to Chemical and Physical Characterization)

* HUSAR
Sulfur in the Atmosphere

STRAUSS
Industrial Gas Cleaning, 2nd edition

THAIN
Monitoring Toxic Gases in the Atmosphere
For Hygiene and Pollution Control

TOMANY
Air Pollution: The Emissions and Ambient Air Quality

* Not available under the Pergamon textbook inspection copy service

ATMOSPHERIC POLLUTION

Its History, Origins and Prevention

BY

A. R. MEETHAM   D. W. BOTTOM

S. CAYTON   A. HENDERSON-SELLERS

D. CHAMBERS

FOURTH (REVISED) EDITION IN S. I. UNITS

Previously entitled
**Atmospheric Pollution
Its Origins and Prevention**

PERGAMON PRESS

OXFORD · NEW YORK · TORONTO · SYDNEY · PARIS · FRANKFURT

U.K.	Pergamon Press Ltd., Headington Hill Hall, Oxford OX3 0BW, England
U.S.A.	Pergamon Press Inc., Maxwell House, Fairview Park, Elmsford, New York 10523, U.S.A.
CANADA	Pergamon of Canada Ltd., Suite 104, 150 Consumers Road, Willowdale, Ontario M2J 1P9, Canada
AUSTRALIA	Pergamon Press (Aust.) Pty. Ltd., P.O. Box 544, Potts Point, N. S. W. 2011, Australia
FRANCE	Pergamon Press SARL, 24 rue des Ecoles, 75240 Paris, Cedex 05, France
FEDERAL REPUBLIC OF GERMANY	Pergamon Press GmbH, 6242 Kronberg-Taunus, Hammerweg 6, Federal Republic of Germany

Copyright © 1981 A. R. Meetham, D. W. Bottom
S. Cayton, A. Henderson-Sellers, D. Chambers

All Rights Reserved. No part of this publication may be reproduced, stored in a retrieval system or transmitted in any form or by any means: electronic, electrostatic, magnetic tape, mechanical, photocopying, recording or otherwise, without permission in writing from the publishers

First edition 1952
Second (revised) edition 1956
Third (revised) edition 1964
Fourth edition 1981

British Library Cataloguing in Publication Data
Atmospheric pollution. — 4th revised ed (S. I. units). — (Pergamon international library).
1. Air — Pollution
I. Meetham, Alfred Roger
614. 7′1   TD 883   79-42852
ISBN 0-08-024003-8 (Hardcover)
ISBN 0-08-024002-X (Flexicover)

Printed in Hungary by Franklin Printing House

Preface to Fourth Edition

"When the London smog of December 1952 was stirring the public conscience in Britain, American cities such as Pittsburg and Los Angeles had already begun to deal with contamination of town air from sources such as domestic backyard incinerators, industrial chimneys and exhaust gases from motor-cars. Now, after further smog incidents in 1956, 1957 and 1962, and a drastic Act of Parliament, there is good evidence that Britain is following suit. Indeed the realization is growing that the British, with their humid maritime climate and dense industrial population, should be following no one, but leading the world in the technique of ridding the atmosphere of its man-made pollution."

This is how the Preface to the Third Edition began. Now, over 20 years after the Clean Air Act of 1956, and indeed 27 years after the senior author produced the first edition, we feel the time is right to present a new edition of "Atmospheric Pollution: its origins and prevention". The three previous editions have, fortuitously, spanned the fascinating history of air pollution control in Britain. The text, initially written to serve as an introduction to air pollution and air pollution control, now stands in its own right as the only authoritative description of our nation's concern with understanding and control of air pollution. Time has produced this metamorphosis from introductory guide to historical review and it is in this light that this new edition is offered and for this reason that the title of the text has now been modified to "Atmospheric Pollution: its history, origins and prevention".

The emphasis throughout the text is upon the evolution of our understanding of sources and controls of air pollution. Since the third edition (published in 1964) enormous strides have been made in the scientific investigation of air pollution; no attempt is made to synopsize these new vast fields of research although reference is made to major current works throughout. However, even a topic as thoroughly investigated as air pollution control does not stand still and the authors feel that in many senses the problems of awareness, monitoring and control of air pollutants have not changed fundamentally from the situation in 1952. Despite Britain's conspicuous success in tackling smoke pollution (described in the third edition) and, more recently, a similar but more limited triumph over sulphur dioxide (described in the National Survey of Air Pollution 1961–1971, Warren Spring Laboratory), there is now the suggestion that particulates (especially lead) and, on a global scale, carbon dioxide and chlorofluorocarbons, may present just as great or greater threats to our environment. Thus the record in this book of earlier approaches and successes may, we suggest, be applied to current problems. The techniques of making surveys of atmospheric pollution; in defining, analysing and explaining pollution levels within geographical areas cannot be divorced from our understanding of dispersion dynamics and meteorological controls of gaseous and particulate constituents.

The interest of the well-educated layman in air pollution and air pollution control has, we feel, also come full circle. The fuel crisis, underlined by the OPEC price increase of 1973, has combined with economic pressures to make well-informed consideration of individual households' fuel use and insulation a sensible, worthwhile and even financially beneficial pursuit.

In keeping with the current trend in scientific literature (and in accordance with the views of the authors), the units used in this new edition will be those of the S.I.—based largely on the metric system. However, since British legislation (especially in boiler practice) is at present still in imperial units these obsolescent units will, where appropriate, be also included in brackets.

Explicit references are gathered at the end of the book together with an extensive and up-to-date bibliography.

We wish to acknowledge the following for supplying data and/or reading specific chapters: Mr. S. C. Mitchell, Department of Fuel and Combustion Science, University of Leeds; Mr. E. F. Curd, Liverpool Polytechnic; Institute of Energy; Solid Fuel Advisory Service; and Mrs. S. E. Mather for her art work.

A. R. MEETHAM, D.Sc., F.R.Met.Soc.
Principal Scientific Officer
National Physical Laboratory

D. W. BOTTOM, F.E.H.A., F.R.S.H.
Former Principal Lecturer
(Environmental Health)
South East London Technical College

S. CAYTON, F.R.S.H., M.Inst.F.
Former Chief Public Health Inspector
County Borough of West Bromwich
John S. Owens Prizeman, 1962

A. HENDERSON-SELLERS, B.Sc., Ph.D.,
F.R.Met.Soc., F.R.A.S.
Lecturer in Geography
University of Liverpool

D. CHAMBERS, B.A., M.Sc., M.E.H.A.
Senior Lecturer (Environmental Health)
School of Biological Sciences
Thames Polytechnic

Contents

1. Introduction   1
    Growth of pollution   3
    Control of pollution   4
    Scope of the book   5

2. Origin of Fuel   6
    Energy and the origin of the Earth   6
        Fossil fuels   7
    Energy value of fuels   8
        Calorific value   8
        Gross and net calorific value   10
        World reserves and annual output   11

3. Natural Solid Fuels   14
    Wood   14
    Wood charcoal   15
    Peat   16
    The coal series   16
        Lignite   18
        Bituminous coal   18
    Ash and sulphur in coal   19
        Washed coal   20
    Coal hazards   21

4. Mineral Oils and Gases   23
    Petroleum   23
    The refining process   24
    Characteristics of fuel oils   28
    Preparation for burning: burners   30
        Vaporizing burners   30
        Pressure jet burners   30
    Natural gas   33

5. Manufactured Fuels   34
    History   34
    Coke   35
    Coal tar and tar oils   38
    Hydrogenation and hydrocarbon synthesis   38

Liquefaction and gasification	39
Alcohol	39
Manufactured gaseous fuels	40
Historical note	42

6. Combustion and Power Generation — 44

Internal combustion engine	47
Atmospheric pollution from engines	48
Cooling towers	50
Electricity	51
Uses of electricity	53
Conclusion	56

7. Industrial Boilers — 57

Coal-fired boilers	57
Vertical boiler	58
Lancashire boiler	59
Economic boiler	60
Thermal storage boiler	61
Water-tube boilers	62
Industrial hot water boilers	64
Boiler instruments	67
Carbon dioxide	69
Smoke as an index of efficiency	71
Alternatives to coal	72
Mechanical stokers	73
Pulverized fuel	76
Boiler availability	78
Soot blowing	79
Fluidized beds	79

8. Industrial Furnaces — 80

Group (1) furnaces	81
Horizontal retorts	82
Coke ovens	83
Vertical retorts	84
Static vertical retorts	85
Electric furnaces	85
Oil refineries	85
Group (2) furnaces	86
Atmospheric pollution from furnaces in Groups (1) and (2)	88
Group (3) furnaces	90
Steel industry	90
Clay industries	90
Lime and cement kilns	92
Atmospheric pollution from furnaces in Group (3)	94
Smoke in the steel industry	94
Sulphur dioxide and grit	94
Summary	95

9. Domestic Heat Services — 96

Choosing a domestic heating system	97
Solid fuel	98
Central heating	100

Gas and electric fires	102
Thermal storage electric heating	102
Thermal insulation	102
Hot water and cooking	104
Coal economy	106

10. Atmospheric Pollution — 108

    Smoke — 109
    Ash — 110
    Sulphur dioxide — 111
        Carbon monoxide and carbon dioxide — 112
        Nitrogen oxides (NO_x) — 113
        Lead, chlorine and fluorine compounds — 113
    Pollution from petroleum products — 114
    Odours — 115
    Radioactive air pollutants — 115
    Pollution from other sources — 117
        Gases from chemical works — 117
        Burning spoilbanks — 119
        Incineration of refuse — 120
        The offensive trades — 121
        Particles — 121

11. Measurement of Atmospheric Pollution — 123

    Measurement of smoke — 124
        Smoke filter — 126
        Self-changing smoke filters — 128
        Portable smoke filters — 128
        Weighable smoke filter — 129
    Measurement of ash and other deposited pollution — 130
        Deposit gauge — 130
        Rapid surveys of deposited matter — 133
    Measurement of sulphur dioxide — 135
        Volumetric estimation of sulphur dioxide — 136
        Portable instruments for sulphur dioxide — 137
        Automatic monitoring — 137
        Sulphur dioxide by the lead dioxide instrument — 138
    Pollution roses — 140
    Microscopic examination of grit — 140
        Microscopic examination of suspended matter — 142
        Other pollutants — 145
    Measurement of daylight — 146
        Use of measurements of atmospheric pollution — 146

12. Distribution of Pollution — 147

    Historical perspective — 147
    Distribution in Britain as a whole — 147
        Deposited matter — 147
        Smoke and sulphur dioxide — 150
    Distribution within a town — 153
        Deposited matter — 153
        Smoke and sulphur dioxide — 153
    Recent surveys — 157

The National Survey of Air Pollution 160
   Smoke 160
   Sulphur dioxide 160
   Concluding remarks 161

13. Variability of Pollution 162

Changes in deposited matter 164
   Yearly cycle 165
Changes in smoke and sulphur dioxide 165
Irregular variation 169
Air pollution meteorology 169
   Chimney plumes 171
Fogs 172
   Constituents of the London fog, December 1952 173
   Quantities 174
   Heat balance 175
   Water balance 175
   Smoke balance 176
   Sulphur balance 176
   Halogens 177
   Oxides of carbon 177
Ground level concentrations 177
   Summary 180

14. Effects of Pollution 181

Biological effects 181
   Health 181
Threshold limit values 184
   Smog disasters 184
   Mortality attributed to smog 185
   Effects on animals 188
   Effects on vegetation 188
Physico-chemical effects 189
   Insulators 189
   Metals 189
   Materials 189
   Fog, visibility and sunlight 191
The cost of pollution 192
   Conclusion 193

15. Prevention of Atmospheric Pollution 194

Prevention of smoke 194
Prevention of ash and grit 195
   Selection of fuel 196
   Design and operation of furnace 196
   Particulate removal 197
   Air conditioning 201
Prevention of sulphur dioxide 201
   Removal of sulphur from fuel 202
   Removing of sulphur dioxide from flue gases 203
   Chimney height 204

16. Air Pollution Control—Law and Administration   207

    The United Kingdom system   207
    The Alkali Inspectorate—structure and responsibilities   208
    Local Government administration and control   209
    The Clean Air Acts   209
    Smoke emissions   210
    The control of chimney heights   211
    Grit and dust from furnaces   212
    Smoke-control areas   213
    Air pollution legislation throughout the world   214
    Air pollution control within the member states of the European Economic Community   214
        Belgium   214
        Denmark   215
        France   215
        Federal Republic of Germany   215
        Ireland   216
        Italy   216
        Luxembourg   216
        Netherlands   216
        United States of America   217
    Motor vehicle pollution   218

    Appendix A.   Conversions   220
    Appendix B.   British Standards   222

BIBLIOGRAPHY   223

INDEX   227

CHAPTER 1

Introduction

THIS book is meant to be of use to all who are professionally interested in atmospheric pollution—environmental health officers, architects, engineers, meteorologists, legislators, city councillors, boiler operators and builders. It is also intended to encourage and help the many citizens who are aware of the degree of atmospheric pollution, and who by their writing, conversation, or example, are helping in the fight against it.

The term pollution is used to describe the admixture of any foreign substance which we dislike with something pleasant or desirable. Atmospheric pollution, therefore, is an undesirable substance mixed with the open air. Any objectionable gas in the air is atmospheric pollution, whether it is harmful or merely unpleasant, but this definition does not apply only to gases. The air frequently contains solid particles or tarry droplets, less than about 10 μm in diameter, which continue in suspension for a long time. Since any particulate matter is liable to cause trouble, these aerosols, as they are sometimes called, are all atmospheric pollution, irrespective of their chemical nature. Also, by general consent, the term atmospheric pollution is applied to larger particles, when these are lifted into the air by the wind or emitted from a chimney. Although the largest particles escape from the atmosphere relatively quickly by falling to the ground, they are capable of causing damage and intense irritation. Finally, there is the risk of radioactive materials in the air, in amounts sufficient to injure public health, whether directly or through selective absorption by food plants and animals.

Some atmospheric pollution, notably that of natural origin, and dust emission from quarries and, more recently, the reintroduction of open-cast mining is produced in the open air. More commonly, the constituents are produced indoors under conditions which are more or less controlled, and are then discharged into the open as the most convenient way to get rid of them. Usually the first thing we do when a room becomes stuffy is to open a door or window; fresh air enters and drives the bad air by some other exit into the open. If pollution is produced very rapidly within a room such natural ventilation may be inadequate, and we equip the room with ventilating hoods, fans, and flue pipes leading to the outside air. Many such installations are to be found in factories where fumes, dusts or gases are liable to be produced; in our own homes every boiler and fireplace has an exhaust flue through which both potential atmospheric pollutants and, perhaps more importantly from an economic point of view, waste heat escape to the atmosphere.

There would seem to be little danger in discharging a small quantity of pollution into the open, where it is so much diluted that it quickly becomes harmless. In 1 sq. km there are about 10,000 tonnes of air below roof level but, when it is realized that as little as one part per million, i.e. 0·01 tonne of pollution in 10,000 tonnes of air, is sufficient to make the air unpleasant to breathe, a simple calculation will show how quickly such a quantity of air may be seriously polluted. Before the Clean Air Acts, domestic combustion sources were commonly responsible for coal consumption at a rate of about 0·5 tonne per minute over every square kilometre of an urban area. On average each tonne of coal produces about 0·05 tonne of pollution, including solid particles and gases such as sulphur dioxide, (but not counting carbon dioxide). When all this pollution was emitted into the atmosphere, as it often was, it was enough to cause serious pollution in the air below roof level in 36 sec. Usually, of course, much of the pollution would spread above roof level and cleaner air would be continually brought in from outside the town and above it; but if this natural ventilation failed, the urban populace would have been gasping for breath in $\frac{1}{2}$ hr, and dead in 5–10 hr.

The success of the Clean Air Act and the switch to other forms of heating sources in preference to coal and smokeless solid fuels has resulted in the replacement of local combustion (domestic coal fires) by centralized combustion (e.g. fossil-fuelled power stations). The total pollution load to the atmosphere is thus not altered, yet the origin of the source is changed. In addition the height of the emission is elevated from roof level (pollution levels compounded by "trapping effects") to several tens or hundreds of metres (power station "tall chimneys").

In the 1950s it was estimated that over 8 million tonnes of atmospheric pollution were produced each year in Great Britain from the combustion of coal and its derived fuels. This caused far more damage than pollution from any other source, even from decaying vegetation, the evaporation of sea spray, or windblown dust. Although current damage due to combustion-derived pollutants is noticeably less it is still important economically and continues to be the most serious source of atmospheric pollution. Pollution from fuel is thus the main subject of this book.

To complete the definition of atmospheric pollution it might seem necessary to enumerate the chemical compounds in it, but this would be an academic exercise of little real value. There is no reason why pollution should not contain any of the naturally occurring chemical elements as well as the man-made or artificial elements (e.g. lawrencium). The various pollution materials which escape from chimneys can be more conveniently divided according to their properties into three groups: (1) the reactive substances, (2) the finest particles, which remain suspended in the air for a long time, ultimately being deposited as dirt on walls, ceilings and other surfaces, and (3) the relatively coarse particles which quickly fall to the ground.

Each group is dominated, as it happens, by one constituent. (1) The reactive substances include sulphur dioxide, sulphur trioxide, carbon monoxide, ammonia, hydrochloric acid, compounds of fluorine and radioactive materials, but sulphur dioxide is by far the most important in ordinary town air. (2) The finest particles are mostly the "smoke" which is produced when fuels are imperfectly burned. (3) The coarser particles are mostly mineral matter and grit from fuel; though smaller in size they are the same material as the ash which collects under a fire, together with particles of unburnt and partly burnt fuel. Much attention will be given to sulphur dioxide, smoke and ash, not only because they are most important but because they serve jointly as prototypes: any other form of pollution when released into the atmosphere will behave similarly to one or other of them.

Growth of pollution

The amount of atmospheric pollution which we endure today is partly the consequence of our living in communities where all kinds of fuel may be burnt, and partly due to our voracious demand for goods manufactured with the help of heat and power which come mostly from fossil fuels. In London, coal was first used on a serious scale in the thirteenth century, when local reserves of firewood were nearing exhaustion. Smoke from this coal soon invoked complaints, and in 1273 Parliament passed an act which prohibited the burning of coal in London. In 1306 an artificer was tried, condemned and executed for this offence. By Elizabeth's time the law had evidently been relaxed, for a deputation of women went to Westminster to see the Queen about "the filthy dangerous poisonous use of coal". Another fruitless effort to rid London of smoke was made by the philanthropist John Evelyn in the reign of Charles II. He proposed that the factories of brewers, dyers, lime-burners, salt and soap-boilers, and others of the same class, should be moved lower down the Thames; and that Central London, as we now know it, should be surrounded by a green belt thickly planted with trees and scented flowers.

In other countries also coal had a limited industrial use. The Italian craftsman Vannoccio Biringuccio wrote in 1540: "Besides trees, stones that occur in many places have the nature of true charcoal; with these the inhabitants of the district work iron and smelt other metals and prepare other stones for making lime for building. But now I do not wish to think of that far-away fuel, for we see that Nature provides things for our every need, and she always generates in abundance of trees".

In England as a whole, and not merely in London, the "abundance of trees" began to be inadequate in the eighteenth century. Before then, goods were manufactured, as the word implies, by hand, and the commonest source of power was the muscular effort of men and animals. There were also water wheels, windmills, and sailing ships but no satisfactory way to convert the energy stored in fuel into mechanical energy had yet been developed. Furnaces were used in a number of industrial processes and, in particular, the iron-workers were heavy consumers of fuel. They required higher temperatures than could be attained by burning raw wood, so they burnt wood charcoal, and though this was an excellent fuel for their purpose, it was obtained by a very wasteful process. The demands of the iron industry, in conjunction with equally wasteful domestic fires and ignorance of the arts of forestry, ultimately caused our national shortage of wood. Moreover, the demands of the iron industry were a major cause of the industrial revolution for men sought a substitute for charcoal, and they discovered how to convert coal into coke which was suitable for metallurgical processes. The coke was mechanically stronger than charcoal, permitting the invention of the blast furnace. This led to an increased production of all forms of iron, and encouraged the invention of engines and machinery made of iron.

In the nineteenth century came the completely unforeseen developments of the industrial revolution. In one manufacture after another, hand-work was replaced by machine-work, and the output of goods increased enormously. Great new industries sprang up in hitherto obscure districts where the requisite raw materials, of which the chief was coal, were found.

New towns arose in a haphazard manner, never keeping pace with the needs of the rapidly increasing population. Life in England was largely changed, ultimately perhaps for the better, since few of us would willingly give up all the advantages that machinery has brought. However that may be, in some ways the change was for the worse, and we are still suffering from the damage undergone by society as a result of those decades of blind economic readjustment. In particular, problems of atmospheric pollution increase very much as a town grows in population density or industrial activity, and for many years the air of most British towns must have rapidly deteriorated.

Our bad habit of polluting the air with waste products of combustion and other chemical processes was formed long ago, but since the beginning of the industrial revolution a minor irritation has become a great social evil. In towns and industrial districts rain water lost its purity; ash and other solids fell continuously to the ground; the air contained a suspension of fine particles which penetrated indoors, to be deposited on walls, ceilings, curtains and furniture; our clothing, our skins, and our lungs have been contaminated; metals corroded, buildings decayed, and textiles wore out; vegetation was stunted; sunlight was lost; our natural resistance to disease was lowered. In a hundred and one ways the miasma of atmospheric pollution lowered our vitality and our enjoyment of life. During the third quarter of the twentieth century public awareness and indeed public outcry against these evils has grown considerably. Citizens are nowadays unwilling to accept that being urban dwellers of necessity implies an increased risk of certain cancers, respiratory illnesses, increased blood lead levels and decreased life expectancy.

Control of pollution

Most progress has been made, so far, against industrial smoke. Smoke is a sure sign that fuel is being wasted, and the manufacturer saves money if he eliminates it by burning his fuel more efficiently. There have been fewer successes against the ash, sulphur dioxide, and other incombustible waste products in industrial flue gases, because there is usually no immediate economic justification for removing them. Since 1933 in America and 1946 in Britain, areas have been marked out in which there are heavy penalties for the emission of smoke, whether from industry or domestic chimneys. Much has already been done to alleviate the pollution in a number of the world's cities. Some advances have been made by individuals and private firms, some by local authorities, and some by central governments. Engineers in Britain, America, France, Germany and other West European countries, Russia and Japan have greatly improved the science of combustion in its application to raw coal, and more efficient combustion has always caused a reduction of smoke. Scientists have developed processes for transforming coal into various smokeless fuels; they have also designed numerous devices for removing dust and sulphur dioxide from flue gases. Manufacturers have tried to keep abreast of advances in engineering and science; their success may be judged by the differences between recent plants and those built 20–50 years ago. Local authorities have fought against pollution with such weapons as they possess, and they have many successes to their credit. The part played by central governments has been to prepare these weapons, by making laws under which local authorities can effectively act, and by sponsoring industrial research. In 1956 the Clean Air Act was introduced following the Beaver Report of 1954. (Some additions and amendments followed in the Act of 1968.) Smoke emissions, and to a lesser degree sulphurous emissions, decreased rapidly. Smoke-control areas were initiated where only smokeless fuels are permitted: London, Manchester and other urban areas were transformed into cities where sunlight amounts were inrceased, incidence of smogs fell and the air quality vastly improved.

Scope of the book

Although atmospheric pollution can be reduced or eliminated in many different ways, each way involves questions of economics, the time factor, availability of materials, priority over other urgent reforms and, to be frank, of individual and social psychology. To provide a basis for consideration of this variety of questions, it is not enough for a book to give information about atmospheric pollution, its measurement, distribution and effects. These matters are important, but there is need to go into the subjects of fuel, fuel-burning applicances, industrial processes, and domestic requirements, not, perhaps, with too technical an approach, but deeply enough to provide at least a foundation for more technical studies.

The next six chapters deal with fuels, furnaces, and fires; the treatment of such subjects has been kept as free as possible from technical jargon. It is necessarily brief, but the reader may find further help in the texts listed in the Bibliography at the end of the book. The eight following chapters are given to a study of the properties of atmospheric pollution. Remedial measures are then considered and the last chapter is an account of the law in Englend and in other countries in so far as it concerns atmospheric pollution.

CHAPTER 2

Origin of Fuel

Energy and the Origin of the Earth

ALMOST all the energy available on the Earth can be ultimately attributed to the Sun. The Earth's surface is exposed to solar radiation (1360 W m^{-2}) which heats both ground and atmosphere, drives the atmospheric circulation and, perhaps most importantly, permits photosynthesis. Hence all the fuels based upon plant and animal by-products as well as the "alternative" energy sources such as solar, wind and wave power depend upon the Sun as a radiant source. Only the nuclear industry, which utilizes spontaneous breakdown of naturally occurring radioactive isotopes (for instance uranium-235) has a power source which is older than the Solar System itself since these heavy nuclides were almost certainly derived from supernovae explosions prior to the condensation of the Sun and its planetary system 4·6 thousand million years ago.

Successful use of available fuel resources depends upon a number of fundamental qualities of the Earth and its environment. In particular, the release of heat by burning (i.e. oxidation) is a satisfactorily and easily implemented procedure since approximately 20 per cent of our atmosphere consists of free molecular oxygen.

The atmosphere of the Earth (and probably all the terrestrial planets) is of secondary origin: that is, not a remnant of the solar nebula. This is deduced from the marked depletion of the noble gases in the Earth's atmosphere relative to the solar abundance, which indicates that any primitive atmosphere must have been very rapidly dispersed after the formation of the planet. An internal source for the atmospheres of all the terrestrial planets has been the generally accepted theory since the comprehensive assessment of possible sources for the "excess volatiles" by Rubey in 1951. Outgassing from the planetary surface would have resulted in a predominantly carbon dioxide atmosphere (similar to the atmospheres of both Mars and Venus today). However, the Earth's atmosphere has been continuously reworked throughout the history of the planet due to the presence of large masses of surface liquid water and life. Indeed the free oxygen in the atmosphere is a direct product of the evolution of plant life on Earth, which, since its origin at least 3·4 thousand million years ago, has produced oxygen as a waste product from photosynthesis—the imbalance between photosynthetic combination of carbon into plant cell tissue and reoxidation and return of carbon dioxide to the atmosphere results from rapid burial of non-decayed organic matter.

Fossil Fuels

The solid framework of plants is made of cellulose, whose molecular formula is $(C_6H_{10}O_5)_n$ where n is an unknown number, probably between 1000 and 5000. Plants produce their entire substance out of carbon dioxide and water, with the help of traces of compounds from the soil which contain nitrogen, phosphorus, potassium and other elements.

It is important to observe that if six molecules of carbon dioxide ($6CO_2$) and five molecules of water ($5H_2O$) were simply joined together, the result would be $C_6H_{10}O_{17}$; thus, in forming cellulose, plants reject twelve out of every seventeen oxygen atoms. To do this they require energy, and they get the necessary energy from solar radiation. Conversely, cellulose may be burnt in air, when it takes up the twelve oxygen atoms, forming water and carbon dioxide, with the emission of heat. The other organic substances in plants are formed in a similar way to cellulose, and they too are combustible.

Plants are among the most fundamental form of life because they utilize carbon dioxide, which is a simple inorganic compound. Animals cannot directly use carbon dioxide; they assimilate their carbon from the plants or other animals which they eat. So even the fuels which come from animals are the indirect consequence of vegetable life, and all the organic fuels, being ultimately derived from plants, owe their potential heat-energy to the light and radiant heat of the sun.

Cellulose is a carbohydrate; that is, besides carbon, it contains atoms of hydrogen and oxygen in the proportion 2 : 1, the same as in water. When carbohydrates are burnt, they yield approximately the same quantity of heat as the carbon in them would produce by itself; their hydrogen is of little value as a fuel because it is already chemically bound. Evidently it is a disadvantage for a fuel to contain oxygen, and it is desirable to consider what other natural organic compounds exist which contain a lower proportion of oxygen than the carbohydrates.

The sugars and lignins found in plants and the proteins that constitute a large part of animal substance are little or no better than cellulose as fuels; but in the metabolism of both plants and animals, some carbohydrate is converted into fats, which are stored as reserves of food within the living organism. Fats contain relatively little oxygen, and when burnt they generate about twice as much heat as an equal weight of carbohydrates. Until comparatively recently, beef and mutton tallow were commonly used as fuels, as well as oils extracted from fish and from the seeds of various plants.

Other fuels are made by bacteria from the remains of dead plants and animals and from animal waste products. The richest natural fuel, marsh gas, is made in this way. It consists mainly of methane, CH_4, which has a double advantage as a fuel since it contains a high proportion of hydrogen but no oxygen.

Today the most important fuels are coal, mineral oil and natural gas and these have been produced by slow geological processes. If dead vegetation undergoes prolonged heating or compression it loses oxygen, and though it also loses hydrogen, it changes into a better fuel. In some measure these changes can be copied in the laboratory. By the action of heat only, wood is converted to charcoal which is nearly pure carbon; by the combined action of heat and pressure, wood or damp cellulose can be converted to a substance rather similar to coal. There is no doubt that coal was evolved in this way by the slow action of heat and pressure, with micro-organisms helping in the process (see Chapter 3). The origin of mineral oil is of a similar nature.

Energy Value of Fuels

Calorific value

It is evident that the usefulness of a fuel must be measured, not merely by the amount available, but by the quantity of heat it will yield. To determine the heat which may be obtained by burning a known weight of fuel, the weight must be multiplied by a factor, dependent on the nature of the fuel. This factor is called the calorific value, and it is simply the number of units of heat obtainable from unit weight of fuel.

The calorific value is measured in specially designed instruments called calorimeters, the details of which depend on whether the fuel is a solid, liquid or gas. The general principle is the same, however, namely to transfer the heat generated to a relatively large mass of water and measure its rise in temperature. A unit of heat is such as will raise the unit mass of water through one degree of temperature. If the unit mass of water is taken as 1 g, and the unit of temperature as the degree Celsius, the corresponding unit of heat is called the calorie which is equal to 4.1868 Joules (J). (The correspondding Imperial unit defined by 1 lb heated by 1° F is the British Thermal Unit (B.t.u.) and is at present (1980) still used in legislation in the U.K.)

Figure 1 is an illustration of a bomb calorimeter, for determining the calorific value of solid and liquid fuels. About 1 g of fuel is burnt in oxygen at a pressure of 25–30 atm, and its heat is taken up by a large mass of water in a calorimeter vessel, the temperature

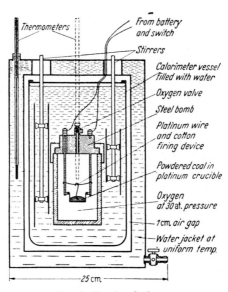

Fig. 1. Bomb calorimeter

rise being accurately measured. The heat taken up by the steel bomb and the calorimeter vessel is allowed for by finding in a separate experiment their water equivalent, i.e. the mass of water which has an equal capacity for heat. Allowance is also made for heat transferred into or out of the calorimeter during the test. Other small corrections are made, as will be seen from Table 1, which gives a specimen determination, somewhat simplified.

TABLE 1. *Determination of calorific value of a solid fuel*
(Bomb calorimeter; Tymawr coal)

	(f)	Weight of coal taken (g)		1·0017 g
		Water equivalent of calorimeter and bomb, etc.	520 g	
		Weight of water used	1980 g	
	(w)	Total weight of water		2500 g
	(T)	Thermometer rise, corrected for cooling of water during test		3·402°C
	(wT)	Heat liberated (2500×3·402)	8505 cal =	35·6 kJ
		Correction for heat from burning thread, used in igniting charge	−36 cal =	−0·15 kJ
		Correction for heat from burning nitrogen from fuel	−12 cal =	−0·05 kJ
		Correction for heat from oxidation of sulphur dioxide and water to sulphuric acid	−25 cal =	−0·10 kJ
$(wT$ corrected)			8432 cal =	35·3 kJ
	(wT/f)	Calorific value, (in J kg^{-1})		35·24 MJ kg^{-1}
		×4·299×10^{-4} = calorific value, B.t.u./lb		15150 B.t.u./lb

Note: Because in the bomb calorimeter the fuel is burnt in oxygen under pressure, reactions take place with nitrogen and sulphur in the fuel which do not occur when the fuel is burnt in the ordinary way. The amounts of the products, nitric acid and sulphuric acid are determined chemically and corrections made for the heat generated by the abnormal reactions. In practice, the correction for sulphuric acid is applied after dividing by f.

(It will be seen that calorific values in J kg^{-1} are converted to B.t.u./lb on multiplying by 4·299×10^{-4}).

Figure 2 is a drawing of a Boys continuous-flow calorimeter for measuring the calorific value of gaseous fuels (of approved method for public gas supplies). The gas is burnt at a steady rate, and water flows at a constant speed through the copper heat-exchanger coil and out through the heavy brass equalizing chamber, leaving the calorimeter at a higher temperature than when it entered. After a steady state has been reached, the test is carried out, three primary measurements being made over a definite interval of time: the volume of gas burnt, the mass of water passed through the calorimeter, and the rise in temperature of the water.

If the calorific value of a gas at a sampling point is changing with time, it would be exacting work indeed for an observer with a simple Boys calorimeter to keep track of the changes and so estimate the average calorific value over a long period. Recording calorimeters have been designed which do this automatically by (1) maintaining a constant ratio between the rate at which gas is burnt and the rate at which water flows,

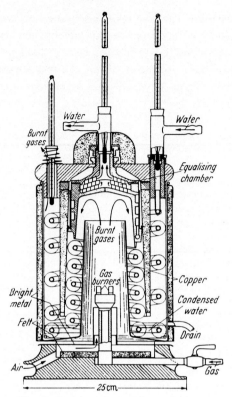

Fig. 2. Boys continuous-flow calorimeter

(2) keeping the ratio air/gas constant, (3) avoiding differences between water vapour content of the emerging products of combustion and the combined gas and air entering the calorimeter, (4) eliminating transfer of heat by radiation, and (5) correcting for deviations from standard conditions of volume measurement. The calorific value is then directly proportional to the temperature rise of the water, and is recorded on a chart. Adiabatic calorimeters, although expensive, are simple to use and are successfully utilised for repetitive and continuous usage.

Gross and net calorific value

When a fuel is burnt which contains either free or combined hydrogen, one of the products of combustion is water vapour, and before the full calorific value can be extracted from the fuel, this water vapour must be allowed to cool and condense as liquid water. If the water vapour escapes, the loss of heat is appreciable, because of the very high latent heat of steam. In calorimetry, there is little difficulty in retaining the water of combustion and allowing it to condense and cool to the end temperature of the calorimeter, and determining the so-called *gross* calorific value. In furnace practice, however, ithe water of combustion nearly always escapes before condensing; so it is of practical importance to know how much heat can be extracted from a fuel when the water of combustion escapes as steam. This is called the *net* calorific value; it is determined by weighing the water produced during combustion, and subtracting its heat of vaporization from the total heat liberated in the calorimeter. In the example quoted in Table 1, 0·046 g of water was formed in the bomb, and the net calorific value was 34·15 MJ kg^{-1} (14,680 B.t.u./lb).

The gross and net calorific values of typical specimens of representative fuels are given in round figures in Table 2.

TABLE 2. *Calorific values of some fuels*

	Gross (B.t.u/lb)	Net (B.t.u/lb)	Net (MJ kg^{-1})
Hydrogen	62,000	52,920	123
Methane	23,715	21,355	50
Natural gas, approx.	21,500	19,800	46
Crude mineral oil, approx.	19,500	19,200	45
Vegetable oils, animal oils, and fats	17,500	17,000	39
Fuel alcohol, approx.	14,000	13,000	30
Coal, approx.	14,500	14,000	33
Carbon	14,600	14,600	34
Coke, approx.	13,200	13,100	30
Lignite, approx.	13,000	10,000	24
Peat, air dried, approx.	10,000	8000	19
Wood, air dried, approx.	9000	7000	16
Sulphur, when burnt to sulphur dioxide	4000	4000	9
Pyrites	3000	3000	7

It will be seen in the above table that for fuels which contain no hydrogen there is no difference between gross and net calorific value. The large differences for lignite and air-dried peat and wood are due partly to the moisture in the fuel, and partly to water of combustion.

World reserves and annual output

Since the origin of life (at least 3·4 thousand million years ago), only a tiny fraction of the total biomass can have helped to form the fossil fuels. In particular much of the fossil fuel reserves we are currently utilizing were derived from the flourishing of life in the Carboniferous. When great quantities of rich fuels began to be needed 150 years ago we began to consume the coal reserves at an alarming rate. Oil has been consumed at a rapidly increasing rate since 1900.

The amounts of fuel produced are very different in different parts of the world, as will be seen from Table 3 which refers to fuel production in the year 1957. Allowance should be made for intercontinental exports of fuel, particularly oil. The table has been assembled from data supplied by those countries, including all the world's most industrial regions, which make reports to the World Power Conference. No part of the figures from U.S.S.R. has been included in either Europe or Asia.

TABLE 3. *World fuel production in 1957 (millions of tonnes)*
(water-power capacity in millions of kilowatts)

	Africa	N. America	S. America	Asia	Europe	Oceania	U.S.S.R.
Coal	41	460	6	220	620	23	330
Oil	3	400	160	200	24	0·3	100
Natural gas	0	270	16	1	6	0	—
Lignite	0	4	0	6	440	11	130
Wood fuel	8	100	0·2	40	50	6	—
Water-power capacity	1	50	4	12	27	2	—

Ten years later, the world production of coal was estimated at 1750 million tonnes, lignite at 1040 million tonnes coal equivalent, oil at 1639 million tonnes and natural gas at 655 million tonnes.

Table 4 will give some idea of the fuel stocks and expenditure in the world as a whole. It has been assembled from various sources and is very inaccurate, but the numerical values are nearly all expressed to one or two significant figures only, multiplied by a power of 10. The world's needs are increasing so rapidly that these estimates for expected lifetimes are ridiculously high. Milton F. Searl, forecasting on behalf of the U.S. Atomic Energy Commission, predicted that the world fossil-fuel con-

TABLE 4. *Approximate world resources of fuel and energy and predicted lifetime*
(1958 and 1975 estimates)

	1958 estimates		1975 estimates			
	Reserves ($\times 10^9$ tonnes)	Years of production at present rates	Present economically recoverable resources		Total ultimate recoverable energy	
			Reserves ($\times 10^9$ tonnes)	Lifetime (years)	Reserves ($\times 10^9$ tonnes)	Lifetime (years)
Solid fuel	3000	1765	640	235	10% recovery 1000	370
					50% recovery 5000	1840
Oil	100	110	130	30	440	100
Shale oil and tar sands	negl.	—	negl.	—	230	—
Natural gas	100	330	95	56	360	210
Uranium	—	—	57	100 non-breeder	153	260
				breeder	9200	15,600
Geothermal	—	—	0·5	200	3	1,000
Non-tidal water power	$6·7 \times 10^{11}$ W	19	27	50	108	200

Units are tonnes of coal equivalent.

sumption in the year 2000 would be over 5 times the consumption in 1958. If consumption had continued to expand at this rate, slightly over 4 per cent per year, supplies would be exhausted in about A.D. 2050.

As time progresses, more reserves may be found, consumption rates (actual and predicted) vary. This is illustrated by the known world reserves in 1975 given in Table 4. For instance, immediately exploitable solid fuel resources in 1975 were estimated at only just over 600×10^9 tonnes compared to known reserves of 3000×10^9 tonnes in 1958. This has led to predictions for lifetimes of these fossil fuels being drastically reduced as a result of increasing consumption rates especially during the sixties.

In the United Kingdom a report by the International Institute for Environment and Development discussed the future of a "low energy strategy" in which standards of living were maintained whilst overall fuel consumption dropped. This situation could be achieved by frugal management and rethinking by both industrial and domestic consumers on energy usage and wastage.

Evidently new sources of power must be provided to supply much of the world's needs into the twenty-first century. There is little doubt that the main contribution could be by nuclear energy, distributed in the form of electric power. However, non-tidal water power might provide twenty times as much electricity as it does at present. If tidal power is included, the figure rises to fifty times as much, and nearly equals our total fuel requirements, but the capital costs render such a hydroelectric project unthinkable.

Political and economic considerations are now becoming almost as important as the reserves themselves. The OPEC oil price increase in 1973 led to a world economic recession whilst more recently Mexico (and possibly Saudi Arabia) have voluntarily reduced their output thereby forcing a reduction in global consumption.

CHAPTER 3

Natural Solid Fuels

Wood

The stems and branches of trees are not homogeneous; they vary considerably in chemical composition and even greater differences exist between different species of tree. The exact composition of wood is therefore quite important when it is used as a source of chemicals such as turpentine or cellulose, and it is important even in considering wood as a fuel. The approximate composition of a typical wood, after being allowed to dry under cover, is given in Table 5.

TABLE 5. *Composition of typical air-dried wood*

	Total %	Oxygen %	Hydrogen %	Carbon %	Others %
Mineral matter	0·5	0·2	0	0	0·3
Water	13	11·5	1·5	0	0
Resin and wax	2·5	0·3	0·2	2	0
Lignin	20	7	1	12	0
Hemi-cellulose	18	9	1	8	0
Cellulose	46	23	3	20	0
Total	100	51	6·7	42	0·3

The value of wood as fire-kindling material depends strongly on its resin and wax. Among different varieties of tree, there are important variations in the percentage of resin and wax: this ranges from 0·7 in beech, for example, to 3·2 in pine.

The table shows that, of the combustible constituents of wood, cellulose and hemi-cellulose contain a high proportion of oxygen, lignin rather less, resin and wax very little; their fuel value increases as their oxygen content diminishes. When wood is burnt the hydrogen, oxygen, and carbon are converted by the action of atmospheric oxygen to water vapour, carbon dioxide and small proportions of carbon monoxide and smoke, all of which may escape into the air. Part of the ash may escape into the air as well, and all the natural moisture of the wood is driven off. As was shown in Table 2 near the end of the last chapter, the heating power of wood is normally about 16 MJ kg^{-1}, its net calorific value, instead of the much higher gross calorific value of approximately 21 MJ kg^{-1}.

Wood fires require a lot of attention, but they are pleasant in the home, and cause less atmospheric pollution than coal but considerably more than modern smokeless solid fuels (see Chapter 5). Inhalation of wood smoke over long periods can have detrimental effects on health and may be responsible for initiating cancerous growths. In 1957 the United States of America used 100 million tonnes of wood per year as fuel and Canada, Finland and India each about 10 million tonnes (see also Table 3). Obviously in Third World countries natural fuels are still used widely. Indeed not only wood but also dung provides a major fuel for combustion. In Kenya 10 per cent of the energy used comes from fuel-wood and charcoal. A new energy strategy based on fuel-wood is aimed at lowering the price of charcoal and reversing the trend towards kerosene —a costly import.

Wood charcoal

Air-dried wood is shown in Table 5 to contain oxygen about 51 per cent by weight, carbon about 42 per cent, hydrogen about 7 per cent, and a little mineral matter. The simplest way to improve its fuel value is by the application of heat in such a manner as to drive off most of the oxygen, when the hydrogen will also be liberated, leaving a residue which we call charcoal. This is more or less pure carbon, according to the thoroughness of the distillation, except that it contains most of the original mineral matter of the wood. Since before 3000 B.C. charcoal was made by piling wood in a heap, nearly covering it with earth, and setting it on fire at the bottom. Some of the wood burnt, providing the heat whereby the rest was converted to charcoal. Though simple, this process was inefficient, for only 30 per cent or less of the original heat of the wood remained in the charcoal.

Modern methods of wood distillation, in retorts heated usually by oil burners, are much more efficient. The gases that are driven off are allowed to condense, yielding turpentine and oils which are more valuable than the residue of charcoal. The best turpentine and oils are produced when the temperature of the retorts is lower than for complete carbonization, but wood has also been carbonized at higher temperatures, to yield purer charcoal and a combustible gas similar to coal gas.

As would be expected, charcoal is a smokeless fuel. It can be burnt safely indoors in braziers provided that all carbon monoxide is burnt at the periphery of the zone of combustion; but in most well-populated countries it costs more than other fuels. In times of petroleum shortage it has been used as the fuel in road vehicles driven by producer gas. It is still useful, also, in certain metallurgical work where a very pure fuel is required. Its chief uses, however, are connected with its remarkable property of adsorbing organic compounds after it has been "activated" or out-gassed by blowing steam through it at 900°C. Activated charcoal, made from specially chosen raw materials such as coconut shells, is used in sugar refining, solvent recovery, and in respirators for protection against poison gas.

Peat

It is fortunate that the micro-organisms which attack dead vegetation do so mainly for the sugars and nitrogenous matter it contains. The result of these circumstances is that, on the whole, dead vegetation loses oxygen and hydrogen more rapidly than carbon, and so tends to higher fuel value the longer it remains.

Peat may be regarded as the lowest grade of naturally improved vegetable fuel. It is derived from the mosses and plants which grow in marshes, the lower parts of whose stems tend to die off while the upper parts go on growing. Because of the water with which both the living and dead growth is sodden, the bacterial decay is much more gradual than that going on in cultivated land or compost heaps, and the woody tissue and fats of the dead plants remain unchanged for a very long time. The time required to form peat beds can be estimated; some of them are as much as 10 m thick, and their thickness is increasing by about 1 m every 300 years.

There are large reserves of peat in Russia, Scandinavia, Germany, the British Isles, Canada and elsewhere in the same latitudes. It can be dug or cut, during summer, and allowed to dry in the wind. When first cut it contains over 90 per cent water, and after air-drying it still contains 20–30 per cent water. In consequence of this fact and of its bulky nature, hand-cut peat is seldom transported far from the peat bog. A modern way of preparing peat is to pump it from the bog and spread it in thin layers to dry. The powdered product may be compressed into briquettes twice as dense as cut peat and much easier and cheaper to transport.

Peat can be distilled to give a kind of coke, as well as tar, ammonia, paraffin, and various oils. Like wood, it makes a noticeable amount of smoke. As boiler fuel, both hand-cut peat and briquettes are satisfactory, and have been used at power stations in Northern Europe. Peat has also been used as a locomotive fuel, but in Ireland, where hand-cut peat was employed during the Second World War in locomotives designed for coal burning, passengers found their journeys were taking them nearly twice as long.

The Coal Series

Although there has been a great deal of recent research into the structure of coals, the picture is not yet complete. It is relatively easy to determine the percentage chemical composition, but the molecular structure is harder to perceive. The formation of coal is due partly to biochemical processes and partly to geochemical processes. Differences in the original organic material lead to different coal composition; Its components are classified as either vitrinite, exinite or inertinite. These classes are known collectively as *macerals*—the main petrographic constituent of British coals being vitrinite (characteristically "bright coal"). Geochemical pressures acting on the decaying organic matter result in a "maturity"— a difference in the degree of coalification or *rank*. The rank increases through the series from lignite to high rank anthracite coals (Table 6). Increasing rank is associated with increasing carbon content and decreasing oxygen and hydrogen content.

The chemical structure of coal can be derived, to a large degree, by an assortment of techniques. Reflectance measurements are important because reflectance is the only

TABLE 6. *The coal series*

	Air-dried fuel		Perfectly dry fuel		
	MJ kg^{-1} net	Volatile matter %	Hydrogen %	Oxygen %	Carbon %
Wood	16		6·5	43	50
Peat	19	50	6·0	32	60
Lignite: brown	24	47	5·5	26	67
Lignite: black	24	41	5·4	19	74
Coal, *bituminous*: long flame, steam and house	25	35	5·0	16	77
Coal, *bituminous*: hard steam, house and manufacturing	32	34	5·0	8	84
Coal, *bituminous*: gas and coking	33	32	5·0	5·4	86
Coal, semi-bituminous	34	11	4·0	2	92
Coal, anthracite	35	8	3·0	2	94

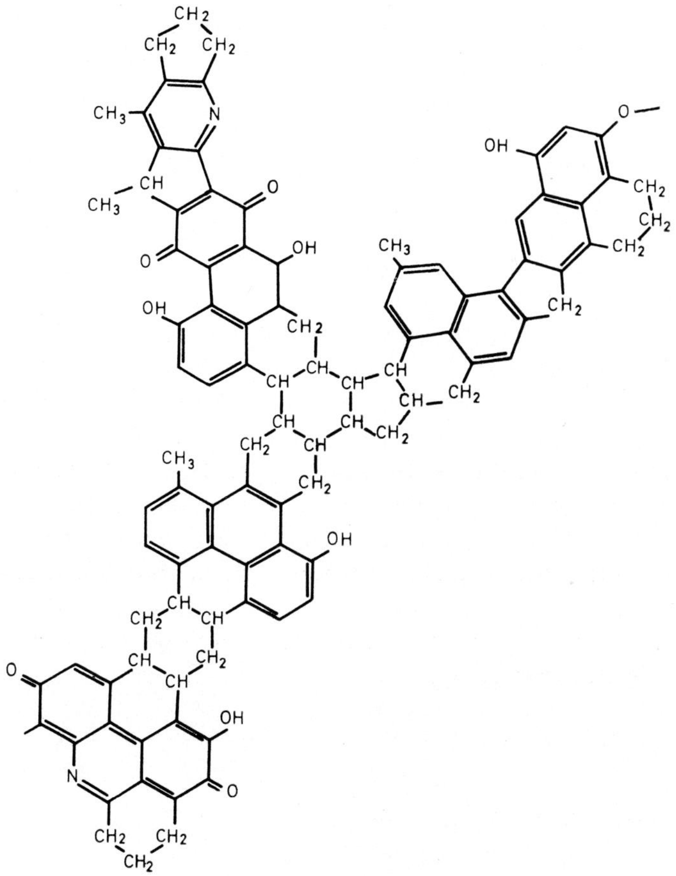

FIG. 3. Proposed molecular model of an 82 per cent carbon vitrinite (after Gibson, 1978)

rank parameter which can be measured using solid coal particles and together with knowledge of the macerals present is used to predict the coking properties of the coal (see Chapter 5). X-ray analysis is often used to deduce the carbon skeletal structure; but the location of the hydrogen and oxygen atoms and other radicals is undertaken using infrared spectrophotometry, nuclear magnetic resonance, ^{13}C resonances (for hydrogen) and infrared spectra of solvent extracts of coal (for oxygen). Nitrogen and sulphur radicals are less easily determined. A typical (agreed) coal structure (for 82 per cent carbon vitrinite) is shown in Fig. 3.

The important properties of coal, from various practical points of view, are (a) its ash and water content, (b) the temperatures of fusion and volatilization of its ash, (c) its proportion of volatile matter, (d) its calorific value, (e) its softening or melting-point, (f) its grindability, (g) its specific gravity, (h) its coherence, (i) its ignition point, (j) its tendency to spontaneous combustion, and (k) its suitability for conversion to coke.

Lignite

Large deposits occur in some areas of a fuel that evidently consists of decayed vegetation, because of its fibrous structure. It is called lignite, from the Latin *lignum*, wood. It may be either black or brown, but even black lignite is easily distinguishable from coal.

Lignite is found in Germany, Australia, New Zealand, Burma, North America, and other countries, including a little in Devon, England. It occurs near the surface in many districts, and can be quarried. It contains a considerable proportion of moisture, and tends to crumble into dust on drying, but it can be compressed into briquettes for transport. It can also be distilled, when it yields a high proportion of tar, but the "coke" is not easily utilized.

Great quantities of lignite briquettes are burnt in Germany, where it is known as "brown coal", in the large closed stoves which, earlier this century, were the principal form of domestic heating on the Continent. The briquettes tend to disintegrate as they lose their last traces of moisture in the upper part of the stove above the combustion zone. The resulting powder is liable to choke the fuel bed, so the German housewives would wrap the briquettes in newspaper before putting them in the stove. As may be expected from its yield of tar and other volatile matter, lignite is far from smokeless. In the larger German towns, domestic flues used to be cleaned twice yearly by the local authority.

Industrially, lignite can be used as boiler fuel, for making producer gas, and for making tar for hydrogenation.

Bituminous coal

The word *bitumen* is derived from the Sanskrit for "pitch-producing". It is usually reserved for the natural pitch-like substances found in the Middle East, Trinidad, and elsewhere. The adjective *bituminous* is applied to coals containing over 20–25 per cent of volatile matter, because of the tarry substances they emit on heating. More than half the economically recoverable reserves of fuel of all kinds consist of bituminous coal. It is found in every continent, and in most large countries of the world.

Anthracite has been named from the Greek word, *anthrax*, meaning coal, but no coal should be classed as anthracite if it contains more than 10 per cent of volatile matter. Anthracite is found in South Wales and Pennsylvania, but in few other places. It is hard, bright and free-burning, its coke being friable and of little value. It is suitable for domestic stoves and boilers, particularly as it is virtually smokeless. Its other important uses are for central heating, steam raising and drying malt and hops.

Types of coal which are intermediate between anthracite and the bituminous coals have been called carbonaceous or Welsh steam coal. They are excellent for all steam-raising purposes, and have been used particularly for marine boilers, although most ships now burn oil. They command a high price; nevertheless if present trends continue they are likely to find increasing application in industrial and domestic fires. Anthracite and other low-volatile coals now have a wide sale in a briquetted form known as "Phurnacite"—see Chapter 5.

Cannel and Boghead are the names given to varieties which are exceptionally rich in volatile matter and so do not fall within the series in Table 6. In some districts they occur as bands within ordinary coal seams. The deposits are relatively small, but may contain from 40 to 90 per cent of volatile matter; thus there may be more volatile matter than in peat or wood. The name "cannel" is supposed to have been given because splinters of it can be lit with a match like candles, and it burns with a long bright flame. Some varieties seem to have been derived from small water-weeds algae; others appear to have originated from the spores of land plants of the fern family.

In the days when gas had to burn with a bright flame, before the invention of the incandescent gas-mantle, cannels and bogheads were much in demand by gas works, because they yield a gas of very high candle-power. In 1920 in Britain, gas became specifiable according to its calorific value instead of its candle-power, and the special properties of cannels and bogheads were no longer required.

Ash and Sulphur in Coal

A small proportion of mineral matter in coal is welcome in specially hot fires, where the ash protects cast-iron fire bars from excessive heat and from corrosive fumes. The ash is also utilized in the chain-grate automatic stoker, where it accumulates at the back of the grate and prevents air from by-passing the combustion zone. In coal required for all other purposes mineral matter is merely a nuisance. It makes no heat, and yet it must be mined, graded, transported, and paid for at coal prices (unless coal is graded and priced according to ash content). It clogs the grate and flues, and if it contains certain substances it may cause frequent stoppages and seriously impair the efficient operation of modern boiler equipment. It lowers combustion efficiency by reducing calorific value, by causing extra opening of fire doors for raking, and by carrying with it, into the ash pit, much useful fuel. If it melts on the fire grate, it becomes even more of a nuisance in the form of clinker. Particles of ash escape up the flues into the open air, and in some districts, until recently, thousands of tonnes per year fell on each square kilometre of ground.

Ash is mainly aluminium silicate, the substance of clay, which softens above 1600°C. If, however, other bases such as ferric oxide and calcium oxide are present, they form double silicates with the aluminium and the temperature of softening is lower. Ash which softens below 1200°C is particularly liable to form clinker. In highly efficient equipment designed to use all possible heat from the coal, even the detailed chemical composition of the ash is important. Examples will be given in Chapter 7, on Industrial Boilers.

Some mineral matter is inherent in coal, having been present in the original vegetation. The rest is derived from the soil which bore the carboniferous plants, and which has been compressed and modified to form rock or shale. At best, there is 1–2 per cent of mineral matter in coal; at worst, fuel containing about 50 per cent of ash can be successfully burnt.

Sulphur, too, is partly inherent in coal and partly extraneous, varying in amount from 0·5 to 4·0 per cent. The extraneous sulphur is mostly in the form of iron pyrites (FeS_2) or mineral sulphates. When coal is burnt or distilled the sulphates mostly remain in the ash; but the pyrites and inherent organic sulphur give sulphur dioxide (SO_2) on combustion, or hydrogen sulphide (H_2S) on distillation.

The chief objections to sulphur in coal are (a) that it may attack the grate bars; (b) that it complicates furnace design in cases where sulphur oxides may not come into contact with the substance being heated; (c) that sulphur dioxide corrodes metal tubes in the flues; (d) that if the gases are allowed to cool below dew point in the flues, sulphuric acid is formed; (e) that hydrogen sulphide in the gaseous fuels derived from coal restricts their usefulness in several ways; and (f) that sulphur dioxide, when released into the air, causes damage and may affect health very seriously.

Oxidation of sulphurous impurities to especially sulphur dioxide is one of the greatest concerns in air pollution. The Clean Air Act of 1956 had a dramatic effect on smoke concentrations, but less so to be but a smaller impact on sulphur dioxide. The pollutant effects of sulphur dioxide are discussed in more detail in Chapter 14.

Coal also contains other substances which are potentially harmful. It contains 0·1–0·7 per cent of chlorine, up to 0·01 per cent of fluorine and small amounts of phosphorus, lead, and arsenic. The study of the effects of releasing these substances into the atmosphere is often hampered by the much greater amounts of sulphur dioxide by which they are always accompanied.

Washed coal

When coal is being graded at the colliery, large pieces of stone can be picked out by hand, and the percentages of pyrites and other mineral matter are thereby reduced. A further reduction occurs if the coal is "washed", and the advantages are so great that a large percentage of British coal is now mechanically cleaned before leaving the colliery.

The specific gravity of coal is about 1·3, of shale about 2·5, and of pyrites over 4·0.

If coal is thrown into a liquid of specific gravity about 1·5–2·0, e.g. a strong solution of calcium chloride or very muddy water, the pieces which are mainly coal will float, while the remainder sink. Or if water is kept in motion through or across a layer of dirty coal, the pieces which contain most coal will move farthest. Similar results can also be achieved by using jets of air in place of water. All these principles are in practical use at collieries.

Washed coal may contain as little as 1 per cent of ash and 0·5 per cent of sulphur, if all the extraneous matter is removed and only the inherent mineral matter and sulphur is left. Although these minimum concentrations are seldom reached by commercial washers, any increase in the proportional output of cleaned coal, or in the efficiency of cleaning, is desirable. It should be noted, however, than an important purpose of coal washeries is to reduce the labour underground of separating coal from minerals, by permitting dirtier coal to be dealt with economically above ground.

Coal hazards

Centuries of hard experience have taught us much general information about coal. As long ago as the early eighteenth century there were mine explosions in which hundreds of miners lost their lives. These explosions may occur if the air in mines contains a high proportion of marsh gas or of coal dust in suspension. With the help of the Davy lamp and other warning devices, and by using inert stone dust as an inhibitor of explosions, the risks have been much reduced; but in spite of careful training, organization, and research, colliery disasters have not been eliminated, though they have become less frequent. In Britain, most underground fatalities are caused by falls of ground or by haulage accidents. Recent years have shown a marked improvement, but the work of increasing safety in mines must go on unremittingly both in the pits and by such organizations as the Safety in Mines Research Establishment of the Department of Energy.

In storage, coal deteriorates, giving off hydrocarbon gases. This used to be of great concern for coal stored at gasworks, where the final yield of gas could be reduced in some cases by about 4 per cent in a fortnight, 8 per cent in 6 months, and 12 per cent in a year. While losing hydrocarbons, coal absorbs oxygen from the air and heat is generated. If the surface area of the coal in store is large in relation to the mass (i.e. if the coal is powdered), and if the heat of oxidation cannot readily escape, the temperature will rise until a faster rate of oxidation begins, and eventually the coal will ignite. As a result of this spontaneous ignition, many fires have broken out in coal heaps and many lives have been lost through bunker fires.

The stagnant air trapped between the fragments of coal in a coal store is a necessary condition for spontaneous combustion to occur. If from the beginning the store is well ventilated by cold air, fire is avoided. Alternatively, water may be used to exclude all air, by storing the coal under water. It should be noted, however, that water in small quantities assists the chemical action, and that very large quantities are needed to put out a fire which has once taken hold of a coal store.

If coal is to be stored with air in the interstices, there are three precautions to be taken against spontaneous combustion. The first is to avoid accumulations of fine coal, such as would normally occur under loading hatches. The second is to provide for adequate ventilation by limiting the height of coal heaps to 3–5 m according to the nature of the coal. The third is to insert iron pipes, closed at the ends, down to the danger points of the store, and take temperatures periodically by lowering in a maximum recording thermometer; increasing temperature is a danger signal, and if the temperature is between 50–80°C fire is inevitable unless preventive action is taken.

In some colliery districts the spoilbanks are often seen and smelt to be smouldering, because of spontaneous combustion. These spoilbanks are dismal enough when they are quiescent, but when they burn they pollute the air and their smoke and sulphur dioxide can do harm to the pastures and crops, as well as to the health of people living near by. The nuisance can be stopped by removing the heap and remaking it in the form of an embankment not more than about 3 m high (see Chapter 10). The application of fine water sprays will not permanently extinguish a burning spoilbank, and the sprays must be kept running continuously.

CHAPTER 4

Mineral Oils and Gases

SINCE prehistoric times a large number of deposits of mineral oils and resins, all of either animal or vegetable origin, have been known to man. Most were practically disregarded, though the Chinese used natural gas as fuel, *amber* has been in constant demand for ornaments, and *bitumen* (asphaltum) was for a long time valued as a natural cement.

In the nineteenth century their usefulness increased: bitumen was needed as a binder for road making, and some of the more easily recovered oils were used for burning and in preparing varnishes. Today, the main value of these minerals is as fuels. Those which are easily converted into the most valuable fuels are being exploited first, but it seems probable that sooner or later we shall be glad to produce fuels, and by-products, from all the richer deposits of mineral oils and resins. At the present time by far the most important of these minerals is petroleum, although a short reference to shale oil is also of interest.

Shale oil of high quality is produced in many countries although it cannot usually compete with petroleum oils without some form of economic protection. However, as petroleum prices increase and reserves become depleted, the exploitation of shale oil seems more likely. It was first produced in Scotland by Dr. James Young, who became known as "Paraffin Young", in the mid-nineteenth century. It was distilled near Edinburgh from certain shales which yielded from 10 to 20 per cent of their weight of oil. The oil was purified by further distillation. In spite of slight chemical differences, shale oil and petroleum have essentially the same uses.

The shale from which some ordinary building bricks are made contains a few per cent of oil, and this burns when the bricks are baked. The heat of combustion is nearly enough to bake the bricks. Thus there is a considerable saving of coal, and the cost of the bricks is appreciably reduced.

Petroleum

The history of petroleum as it is known today begins in 1859 when Colonel Drake drilled his first well in Pennsylvania in the United States of America. Petroleum gets its name from the Greek *petra* meaning rock, and the Latin *oleum* which means oil. It originates from various forms of organic life.

If a rock formation is imagined in which a dome or anticline of impervious rock covers a region of porous rock (see Fig. 4), and if the latter is saturated with water containing oily organic matter, in the course of a very long time the oil globules may be expected to rise and collect in the covering dome. In this way disseminated oil would become a potential oilfield.

To begin an exploration for oil, the geologists and geophysicists study the surface indications and the geology of an area, trying to get a clear idea of the subsurface structure. They use aerial and satellite surveys, seismographic records and any other available evidence. They cannot detect oil directly; what they do is locate structures of impermeable rock under which, if oil exists, it is likely to have accumulated. The

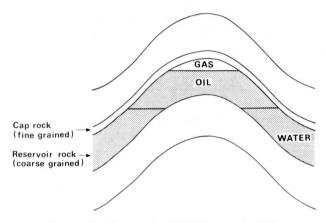

FIG. 4. An oil trap (after McMullan et al., 1976)

proof can only come by the drill, and it is possible with modern methods to explore down to great depths below the surface of the ground. When oil is located a few wells can be judiciously sited and the first phase is complete.

Discoveries of oil and natural gas in the North Sea (see Fig. 5) have allowed Britain to decrease her oil imports from the major oil-producing countries; especially those in the Middle East. America's oil resources are largely in Alaska and the Gulf of Mexico, but are still insufficient to meet all her requirements. The U.S.S.R. is on the whole self-sufficient.

The Refining Process

Crude oil as won from the ground is hardly ever directly usable and it must go through one or more manufacturing processes, collectively called refining, to convert it into the various products required by the markets. Since crude oils from different sources have widely varying characteristics, the refinery must be capable of enough flexibility to cater for these variations. Though formerly the refineries were all near the oilfields and the refined products were transported to the consumer countries, in recent years the practice has been to ship the crude oil to refineries located in the consumer countries. An illustration of such a refinery is shown in Fig. 6.

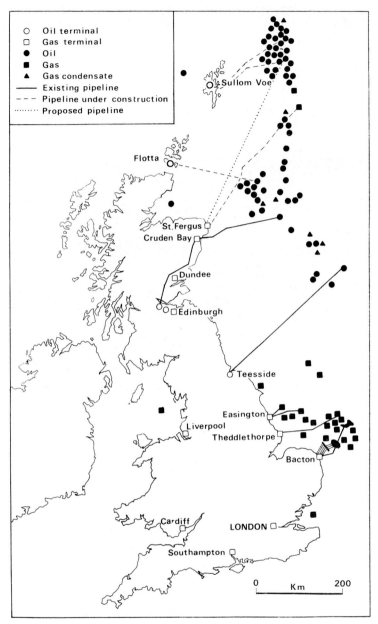

FIG. 5. Oil and gas fields: 31 Oct. 1977. (Reproduced with permission from a map (United Kingdom) produced by the Department of the Environment)

Crude petroleum is predominantly a mixture of hydrocarbons: the typical range of values for the main elements present is: carbon 85–90 per cent, hydrogen 10–14 per cent, oxygen 0·06–0·4 per cent, nitrogen 0·01–0·9 per cent and sulphur 0·1–7 per cent. There are many thousands of separate and distinct hydrocarbons contained in crude oil, and except for the few lightest ones, the difference between the boiling-points of one com-

Fig. 6. General view of the British Petroleum Company's Kent Oil Refinery showing: left and right background the Power Station and Vacuum Distillation Units; and right foreground Propane-Deasphalting Tankage

pound and its neighbour is only a fraction of a degree. They fall into two chemical classes: the paraffins or carbon-chain compounds and various carbon-ring compounds (the aromatics and cycloparaffins). To meet market demands, which vary widely, the oil refiners make use of various processes, the more important of which can be grouped as follows:

1. *Separation processes (Distillation, Absorption, Solvent extraction)*

These depend on differences in physical properties of the hydrocarbons and produce no chemical changes. Distillation and absorption separate the hydrocarbons chiefly according to the size of the molecules, while solvent extraction is useful in separating paraffins from aromatics.

Figure 7 shows an illustration of a simple distillation plant. (Separation is usually undertaken by this method.)

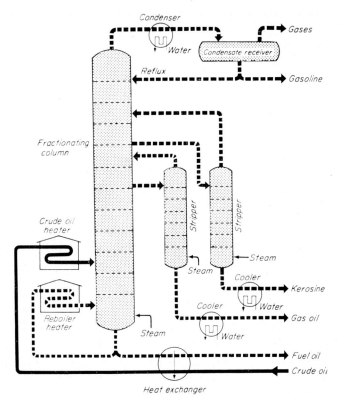

Fig. 7. Simple distillation plant

2. *Conversion process*

In order that the market demand for products can be met, it is necessary partly to change the chemical structure of the hydrocarbons. The main processes employed are catalytic, hydro- and thermal cracking, in which the hydrocarbon compounds are broken down. In all these processes both temperature and pressure control are important. In addition larger and more complicated molecules may be built up from light gases by polymerization, alkylation and isomerisation. Catalytic reforming can be used to upgrade feedstocks.

3. *Treating processes*

The products have usually to undergo a treating process which either removes impurities or changes them into a harmless form. (Removal of sulphur is the most important.) Samples are taken of all products and checked in the laboratory to ensure that the quality is constant.

Blending

A large number of refined products is produced by blending the "base stocks" to give different octane ratings, viscosities, specific gravities, etc.

Characteristics of Fuel Oils

From the wide range of products manufactured in a modern oil refinery only those which are sold as fuel oil will now be discussed. The following properties of fuel oils have to be specified so that the purchaser can use them effectively.

1. *Calorific value.* The calorific value of an oil is the amount of heat given off by unit quantity when it is completely burned. This is normally expressed in joules per kilogram ($J \cdot kg^{-1}$)—usually as a number of megajoules (MJ) per kilogram (1 MJ = 10^6 J).

2. *Viscosity.* The viscosity of any fluid is a measure of its resistance to flow and is defined as the force needed to shear a cube of the fluid of unit dimensions at unit speed. The S.I. units of kinematic viscosity are $m^2 s^{-1}$. However, the "Stoke" is more usually used (1 $m^2 s^{-1}$ = 10,000 Stokes). There are at present in use, however, for assessing the viscosity of oils a number of empirical methods due to Redwood, Saybolt, Engler and others, differing in different countries. The Redwood instrument, still widely used in the United Kingdom, measures the time taken for a standard volume of oil to run from an orifice of standard size, the oil being maintained at the particular testing temperature of 38°C. Viscosities are expressed in seconds, the "thicker" or more viscous the oil, the longer the requisite quantity takes to pass through the orifice.

3. *Specific gravity.* This is defined as the ratio of the weight of a given volume of the oil to the weight of an equal volume of water. Since oil expands on heating and contracts on cooling, specific gravities are usually measured at a standard temperature. In Britain this is 16°C. (Fuel oils are for the most part sold on a volume basis, i.e. at a price per litre).

4. *Flash point.* The closed flash point is the temperature at which the oil produces sufficient vapour in the closed space of the flash point apparatus to ignite when a flame is introduced. Flash point has no bearing on performance, but a minimum is laid down to ensure safety in storage. This minimum may be 66 or 38°C according to the grade.

5. *Pour point.* This is the lowest temperature at which the oil will just flow under the conditions of a laboratory test and it has an important bearing on the storage and handling of the heavier grades of fuel oil.

6. *Sulphur content.* Different oil fuels contain from 0 to 4 per cent sulphur. Those derived from the residual oils of the distillation process generally contain the most.

7. *Ash content.* The lighter grades of oil contain virtually no ash; the heavier grades may contain by specification up to a maximum of 0·2 per cent but in practice this is generally below 0·1 per cent.

8. *Water and sediment.* The water and sediment content of the light distillate grades of oil is extremely small but the heavier, more viscous oils tend to keep sediment and water droplets in fine suspension. Generally in fuel oil the maximum is specified as 1 per cent but the total water and sediment is usually below 0·5 per cent in practice.

Mineral Oils and Gases

TABLE 7

	Class C	Class D	Class E	Class F	Class G
Viscosity—sec. Redwood 1 at 38°C	—	35	250	1000	3500
Gross calorific value MJ kg^{-1}	46·4	45·5	43·4	42·9	42·5
Specific gravity	0·79	0·835	0·932	0·95	0·97

Notes. (1) (Class C is kerosene. Its viscosity is too small for measurement in the Redwood system, but it is approximately 2 centistokes at 15·6 °C.
(2) Classes D, E, F, G are described in British Standard Specification B.S. 2869.

Table 7 gives some of the properties of five grades of fuel oil and the following is a general outline of their uses.

Kerosene (light refined distillate domestic oil)—a grade which has been produced primarily for the flued-type of domestic appliance which incorporates the perforated ring, blue flame burner operating on natural draught.

Gas oil also a light distillate fuel known as domestic fuel oil. This oil was first used in the gas industry for enriching water gas, hence its name. It is widely used for firing heating and hot water boilers up to a capacity of 150 kW, and also finds many other uses in industry, e.g. gas oil is used for many drying processes. It is suitable for use with forced draught, pot type vaporizing burners and also finds extensive use with fully automatic pressure jet burners. It is sometimes referred to as "35 sec oil" (see Table 7, Class D).

Light fuel oil (Class E. Viscosity 250 sec Redwood No. 1) is greatly used for firing the larger sectional hot water and heating boilers. Amongst other applications for which it is used are baking ovens, smaller steam boilers and small furnaces.

Medium fuel oil (Class F. Maximum viscosity 1000 sec Redwood No. 1) finds many uses in larger steam boiler plant and numerous industrial furnace applications. Since 1920 the coal for firing ships' boilers has been very largely replaced by oil.

Heavy fuel oil or *bunker oil* (Class G. Maximum viscosity 3500 sec Redwood No. 1) was designed for the very large steam boiler plant and furnace work where the consumption would justify the additional cost of handling. Its sulphur content is high, over 3·5 per cent, and the flue gases carry away 7 tonnes of sulphur dioxide for every 100 tonnes of fuel burnt.

The grades E, F and G, which contain proportions of residual oils, must be heated either in storage or in passage from storage:

(1) to ensure that the oil will be sufficiently fluid to flow from the storage tank to the transfer pump at the desired rate under the combined influence of atmospheric pressure and head of oil in the tank;
(2) to limit the power required by the pumping equipment; oil may be pumped at viscosities of 20,000 sec Redwood No. 1, but there is an appreciable saving of power consumption at 5000 sec Redwood No. 1;
(3) to prevent stationary oil in pipelines from becoming extremely viscous;
(4) to deliver the oil to the burners at the temperature required for efficient atomization.

The temperatures should be as follows: for Class E, 7°C; for Class F, 18°C in storage tank, suction lines and pumps, and 27°C in ring main; for Class G, 24°C in storage tank, and 38°C in suction line, pumps and ring main.

The heating may be by low pressure steam coils or by a thermostatically controlled electric immersion heater fitted to the tank below the level of the draw-off connection. Outflow heaters, designed to heat the oil as it is required by the burners can similarly be either steam or electric. Where the suction lines between the tank and pumping unit are in an exposed position heating may be provided by means of small bore tracer pipes which are attached to the oil line, and the whole assembly covered by lagging. Alternatively, electric heating cable can be used in a similar manner.

Preparation for Burning: Burners

If one attempted to introduce a thick stream of fuel oil into a furnace and burn it, the outer layers would probably burn; but the central core, having insufficient air, would not burn completely and large quantities of smoke would be formed. The oil must either be vaporized or broken up into a fine mist of small droplets ("atomized") in order to expose as great a surface as possible to air.

The function of an oil burner is therefore firstly to prepare the oil for combustion, secondly to introduce it into the combustion chamber at the rate necessary to provide the heat release required, and thirdly to control the air supply necessary for combustion. Optimization of the combustion process is accomplished by careful consideration of the *time* of contact between fuel and air, the *temperature* of operation and the degree of *turbulence* to ensure adequate mixing. (These three parameters are usually referred to as the "Three T's".) Means of ignition and appropriate safety devices are also needed. Some important types of burner are now considered.

Vaporizing burners

The lightest petroleum fractions, propane and butane used for domestic cooking, vaporize readily. Heavier fractions (kerosene or paraffin used especially for domestic central heating systems) are vaporized by evaporation from a simple wick. Nowadays many central heating systems use the more complicated, fan-assisted, electrical ignition *wall-flame* burner (see Fig. 8).

Pressure jet burners

All oils more viscous than 35 sec Redwood at 38°C have to be atomized. Such oils are prepared for combustion by driving them under pressure through an atomizing nozzle and projecting the cloud of atomized droplets into the combustion chamber mixed with the right proportion of air. The nozzle shown in Fig. 9 both atomizes the oil and controls the rate of oil flow and width of flame. It consists of a conical swirl chamber (B) into which the oil is introduced by means of two or more tangential ducts

Mineral Oils and Gases

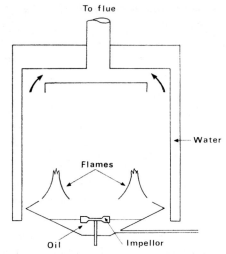

FIG. 8. Wall flame burner (after McMullan *et al.*, 1976)

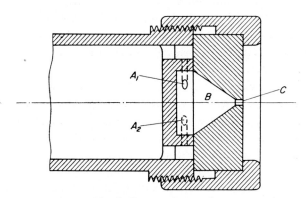

FIG. 9. Pressure jet nozzle

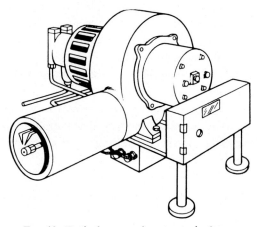

FIG. 10. Typical automatic pressure jet burner

(A) and the oil rapidly rotates towards the apex where it is projected through the final orifice (C) taking the form of a hollow column rapidly rotating about a cone of air.

A typical fully automatic pressure jet burner is shown in Fig. 10. Where it is intended to use an oil requiring pre-heating the burner has an electric element, thermostatically controlled by line heater incorporated into the construction, the purpose of which is to raise the temperature of the oil to that recommended for atomization.

One example of efforts to improve the efficiency of the combustion of liquid fuels is the "high intensity combustion unit" in which atomized oil is burned in a specially shaped refractory-lined combustion chamber. Secondary air is introduced in such a way as to promote a high degree of turbulence and consequent intimate mixing with the atomized particles of oil, and this together with recirculation of hot gases into the root of the flame greatly accelerates the rate of combustion, and in comparison with conventional oil burners, a much greater heat release is possible within a small combustion space. The products of combustion emerge from the unit in the form of a stream of high temperature, high velocity relatively inert gases, which are used directly to heat the specimens. This type of combustion is an advantage, for example in the heat treatment, forging and hot pressing of metals since the inert atmosphere is advantageous in minimizing metal loss due to oxidation. In another application in galvanizing baths, the high-intensity combustion units supply a more even heat than a naked flame and reduce the possibility of local overheating which is one of the main contributory factors in reducing the life of the bath.

Another important development is the "Oil Gasifier". This is designed to produce products of partial, or incomplete combustion, although these products are not in fact a true gas. A fairly small amount of the heat energy is utilized to effect vaporization and partial combustion with a deficiency of air within the gasifier. The resultant products of partial combustion are then ducted (not too far away) to the ultimate point of use where combustion is completed. Gasifiers have been applied mainly to furnaces for metal melting, metal reheating and malleableizing; also for lime kilns.

A further specialized development has been the introduction of the impulse injection system for firing continuous-chamber brick kilns. With this system a metered quantity of oil is injected at controlled intervals into the kiln from injectors arranged along the top or side of the chamber being fired. The equipment is made to be portable so that it can be moved from chamber to chamber as required. Injection is the in form of a "slub" of oil and no attempt is made to atomize it, the oil breaking up when it impinges upon a "striking point" located inside the combustion zone. The air for combustion is admitted into a remote section of the kiln where fired bricks have been previously withdrawn and thus it is pre-heated during its passage towards the chamber which is being fired. This type of burner is suitable for this specialized application, as it provides long lazy flames which are necessary to secure an even distribution of heat throughout the chambers of the kiln.

Natural Gas

This name is given to the combustible gases which, in some districts, are recovered from underground. Not all natural gas occurs in association with oil accumulations. A combustible natural gas occurs in association with coal and, in theory, at any rate, appreciable quantities could be recovered in the coal fields. Even in Britain, however, where the potential demand is great, production has only been on a small scale and only for experimental purposes.

Natural gas contains from 50 to 99 per cent by volume of methane (CH_4), and differing amounts of higher paraffins, together with some inert gases, mainly carbon dioxide and nitrogen. The calorific value may vary between 25 and 60 MJ m^{-3} at standard temperature and pressure.

The term "wet gas" is used to describe a natural gas which contains pentane (C_5H_{12}) (boiling-point 27·9°C) and heavier paraffins with still higher boiling-points at ordinary atmospheric pressure. "Dry gas" is that which contains only a slight amount of these hydrocarbons. Natural gasolne is obtained from wet gas by such methods as compression and cooling; absorption; absorption and low temperature separation. "Liquified petroleum gas" is the fraction of wet gas, mainly butane and isobutane (C_4H_{10}, normal boiling-point –0·5°C and –11·7°C), which can safely be transported in the liquid state under pressure in steel cylinders.

Natural gas can be used for practically any heating purpose and is easily distributed through a system of pressure mains, but in oil fields far away from industrial regions the gas is still largely burnt to waste quite apart from the loss at new wells before the flow is controlled. Natural gas is used in many countries. Over 300 Gm^3 were produced in the U.S.A. in 1956. In 1972 this figure had risen to 655 Gm^3.

In Britain methane was imported in special tankers in 1963. The gas was liquified and carried in very well-insulated tanks, kept below – 160°C by a refrigerating system. A year later, the gas fields of the North Sea began to produce gas with a very low sulphur content, and by 1972 the consumption of North Sea gas had risen to 59 Gm^3.

CHAPTER 5

Manufactured Fuels

History

With one important exception, no attempt was made for thousands of years to modify or improve on the fuels of nature. The exception was charcoal (see Chapter 3) which has been used since prehistoric times for the smelting of metals and other special purposes. Yet natural fuels, particularly coal, usually fall far short of the ideal. Coal is such a complex mixture that to burn it properly requires much skill and attention. In addition it is difficult to store and transport; its ash has to be disposed of; and its products of combustion have to be mixed with many times their volume of air before they are fit to breathe. Thus there is an incentive to transform it into fuels which can be burnt with little handling or other attention in simple appliances, and whose products of combustion are as nearly as possible innocuous.

There is also the possibility of starting completely afresh with industrial wastes or vegetable raw materials, and synthesizing fuels of the required properties. Coal continues to be used mainly because it is in general cheaper and more abundant than artificial fuels, and because very much labour and vast capital equipment would be needed to produce the requisite quantities of alternative fuels. Table 8 shows some of the more

TABLE 8. *Artificial fuels*

3000 B.C. (or earlier) Charcoal	
1587 A.D. Earliest references to coke making for malting etc.	
1709 A.D. Metallurgical coke	ABRAHAM DARBY (Britain)
1780 A.D. Water gas	F. FONTANA (Italy)
1792 A.D. Coal gas, with coke, tar and other products	WILLIAM MURDOCH (Britain)
1830 A.D. (approx.) Producer gas	BISCHOF (Germany)
1833 A.D. Commercial alcohol, from vegetable matter, used as illuminant	U.S.A., France, and Britain
1873 A.D. Carburetted water gas	T. S. C. LOWE (U.S.A)
1907 A.D. Low-temperature coke	T. PARKER (Britain)
1918 A.D. Liquid fuels from hydrogenation of coal and tar	F. BERGIUS (Germany)
1920 A.D. Sewage-sludge gas	H. PRUSS (Germany)
1925 A.D. Synthesis of methanie and liquid fuels from carbon monoxide and hydrogen	F. FISCHER and H. TROPSCH (Germany)
1950 A.D. (approx.) Enriched plutonium—nuclear fuel	U.S.A./Britain
1977 A.D. Water hyacinth farming in open lakes for combustion	U.S.A.

important artificial fuels, and it is worth noting how recently most of them have come into commercial use.

These artificial fuels (except charcoal) are briefly discussed below in the order solids, liquids, gases.

Coke

If any member of the coal series is strongly heated in a vessel from which air is excluded it gives off combustible gases and leaves a residue of which a high proportion is carbon. The process is called "carbonization". Wood and peat yield coherent residues which may be used as lump fuel. Lignite and the bituminous coals of high oxygen content leave residues which easily crumble into powder; so do anthracite and carbonaceous ("steam") coal.

A number of the bituminous coals of low oxygen content behave quite differently. On heating they first appear to melt; the bubbles of volatile matter escaping from the viscous, semi-molten mass cause it to swell, in some cases to several times its original volume; and on further heating the mass hardens and takes the familiar spongy appearance of coke, which it retains when it cools to a normal temperature. These are the "strongly caking" coals which are among the most useful for the manufacture of gas and coke; all sizes down to the finest can be used since the component lumps and particles fuse together.

The majority of coke production today (especially in the United Kingdom) is to fuel the iron and steel industry. The coke required by these industries is often known as metallurgical coke (see Fig. 11). World coke production in 1972 is given in Table 9.

FIG. 11. Metallurgical coke

TABLE 9. *World coke production in 1972 (millions of tonnes)*

W. Europe	U.S.S.R.	U.S.A.	Japan	Total
85·7*	79·8	58·8	36·2	340

Note: In the U.K. in 1972/3, $15·0 \times 10^6$ tonnes of coke was produced by the carbonization of $22·8 \times 10^6$ tonnes of coal.

Metallurgical coke of the highest quality is made from strongly caking coals of 20–30 per cent volatile matter; less strongly caking coals of 30–35 per cent volatile matter are also used. The coal is carbonized at 1000°C or more in coke ovens (see Chapter 8). In 1957, a peak year, Great Britain produced about 20 million tonnes of metallurgical coke, and the U.S.A. about 69 million tonnes.

The most important use of metallurgical coke is for smelting iron in the blast furnace. Besides being hard and strong, the coke should contain as little sulphur as possible and be relatively free from ash, since both must be removed in the slag, requiring extra lime as flux and additional heat for melting and reducing the coke output of the furnace.

Coke for smelting was originally made in heaps similar to charcoal heaps; large coal was stacked round a central chimney, covered with small coal or coke breeze, and kept burning for about 10 days. Later, ovens were built of brick in the shape of beehives, and successive charges of coal were carbonized in them in periods of approximately 3 days; the heat for carbonization was still obtained by burning some of the coal in the charge. Some beehive ovens are still in use, but the modern method is to carbonize the coal in ovens lined with silica brick, heated externally, and to recover tar and ammonia as by-products. Some coke ovens are heated by burning the gases distilled from the coal and surplus gas is used about the works. In districts where coke is made from coal of 30–35 per cent volatile matter many of the ovens are heated by blast-furnace gas, and the richer coke-oven gas is purified and sold. In former years large quantities of coke-oven gas were wasted and even in 1957 over 2 per cent of the gas made in Britain was unaccounted for.

Gas coke (see Fig. 12) was the term given to coke derived (as a by-product) from the production of coal (or town) gas. The demand for gas coke as a smokeless fuel grew

FIG. 12. Gas coke

rapidly as Clean Air Zones were introduced. However, the replacement of town gas production (from coal and later from oil) by North Sea gas in the United Kingdom resulted in the removal of gas coke from the market.

In the financial year 1958–9 the production of gas coke for sale of all sizes was 10·9 million tonnes, and in addition 2·5 million tonnes of breeze (i.e. coke below 1 cm in size) was available for sale. Now no longer are cokes such as Cleanglow, Phimax, Gloco and Sebrite in common use.

Some cokes do, however, remain on sale, but these must now be made specifically for the solid fuel market and must comply with the specifications of the Clean Air Acts of 1956 and 1968.

Sunbrite. This is the trade name adopted by the National Coal Board for their hard cokes in the domestic size range. It is a smokeless fuel produced in coke ovens from a blend of carefully selected coal and is a very consistent high-grade fuel which is most suitable for domestic stoves, cookers, and central heating boilers. It is a firm compact fuel which stores well and because of its greater bulk density occupies less storage space than gas coke and on burning yields a fine ash which passes easily through the fire bars. To obtain the best results it is important that the correct size is used.

Warmco. This is a smokeless fuel produced by the National Coal Board from a clean graded coal at their coke ovens at the East Midlands Division. This fuel is more reactive than hard coke and is suitable for use on an open fire as well as in the closed domestic appliances.

Low-temperature coke (Fig. 13) is made usually from medium-caking or weakly-caking bituminous coal, in "nut" size or "smalls", carbonized at under 650°C. It contains about

Fig. 13. Low-temperature coke

10 per cent volatile matter, and ignites and burns easily; it is suitable for domestic use. Considerable research on low-temperature coke was done at the Fuel Research Station, Greenwich, England (now Warren Spring Laboratory, Stevenage), in the 1920s, and it was found to radiate more of its heat than coal when burnt in an open fire. In similar conditions it radiated 25 per cent of its heat and coal radiated only 22 per cent.

The best known of the low-temperature cokes are distributed nationally under the trade names "Coalite" and "Rexco". These are in great demand, both within Smoke-control Areas and outside because they will burn easily and smokelessly without the necessity of adapting fireplaces.

All kinds of coke are practically smokeless, however they are burnt. Contrary to a common belief, the proportion of sulphur in coke is no higher than in the coal from which it comes; the smell of sulphur dioxide is not particularly noticeable in the flue gases from a coal fire, because of the smoke with which it is mixed. Unavoidably, however, there is a higher proportion of inherent ash in coke than in coal—up to half as much again.

Processed coals made in South Wales from selected low-ash Welsh dry steam coals are carbonized ovoids. The coal is ground to a small size, mixed with a pitch binder, moulded and finally carbonized in a retort. The resulting fuel is uniform in shape, has a low-ash content and has very similar burning properties to anthracite.

These ovoids are sold as "Phurnacite" and are distinguishable from other ovoids, which are not smokeless, by two grooved rings on the outside.

Homefire is a charred briquetted fuel recognizable by its octagonal shape manufactured by the National Coal Board to meet the new demand for a good-quality smokeless fuel. It is easily lit, burns brightly, recovers quickly from slow burning and is much denser than gas coke which is light and bulky.

Coal Tar and Tar Oils

In gas works, about 5 per cent of the original coal was recovered as crude tar, and smaller amounts in coke ovens. Tar was used as a liquid fuel in furnaces, if its most volatile constituents were removed by distillation until its flash point rose above 66°C. Care was needed to prevent the original tar from containing much free carbon or water, either of which could choke the jets through which it was forced into the combustion chamber. However, tar was not normally burnt in its nearly crude state, for some of the substances in it were too valuable to burn.

Usually, tar was carefully divided into fractions by distillation and the light oils driven off below 170°C were redistilled to produce motor benzol. The residual tar was separated into solvent naphtha and heavy naphtha; anthracene oil; creosote (light for disinfectants—heavy for diesel fuel, boiler fuel, or preservative of wood or iron); solid aromatic hydrocarbons; pitch; and road tar, which is pitch dissolved in the heavier oils of the original tar.

Hydrogenation and Hydrocarbon Synthesis

One way of increasing the calorific value of a fuel, as we have seen, is to remove oxygen from it. A still more effective way is to add hydrogen to it. At high temperatures, under pressure, hydrogen readily combines with carbon compounds, producing a complex mixture of hydrocarbons which can be separated by distillation. The proportion of the various hydrocarbons can be altered by varying the temperature and pressure in the reaction chamber, and the chemical reaction can be speeded up by the choice of a suitable catalyst.

Fuel oils, and other products, are now being made by inducing hydrogen to combine with coal, creosote, tar, or carbon monoxide. The rapid strides of the hydrogenation and synthesis industry were due to the strategic need of countries with inadequate natural reserves of oil to have domestically produced supplies available in time of war. During World War II it was estimated that Germany was producing 5 million tonnes per year of liquid products by various methods, but these were costly in comparison with natural-oil products. As might be expected, the chemical techniques are too elaborate and highly specialized to be discussed in this book.

Liquefaction and Gasification

Liquefaction of coal, to produce liquid fuels, analogous to those produced from oil, is still uneconomic in many countries. However, there has recently been considerable success reported by South African research groups. Degradation of the complex coal structure can give similar liquids to crude oils and the term Syncrude (from the oil industry) has often been misapplied to these coal-derived products. Synthesis via gasification can also lead to liquid fractions.

Gasification of coal is commonly done to provide a fuel which will combust more readily. This can be undertaken by use of steam or hydrogen—the products possessing different attributes. Catalytic gasification is currently under development by EXXON.

Alcohol

Alcohol is a manufactured fuel which, mixed with gasoline, has been used in internal combustion engines of the ordinary car type. From the viewpoint of world economy, alcohol is interesting because it can be derived from solar energy in a few months, compared with the millions of years required by all other fuels except wood and, perhaps, peat. It is made by fermentation of vegetable matter—often wasl-e such as molasses, surplus grain, potatoes, or wood—but sometimes grown for the purpose. Alternatively it can be synthesized from ethylene or acetylene, which can be made with electric power from coal and limestone or similar raw materials.

The chemical formula of ethyl alcohol is C_2H_5OH; its composition, by weight, is carbon 52 per cent, oxygen 35 per cent, hydrogen 13 per cent. Compared with petrol, alcohol has a lower calorific value because of the oxygen it contains. Yet because it will tolerate a higher compression without pre-ignition, it is in theory nearly as efficient as petrol in internal combustion engines. In practice it presents difficulties because it absorbs water vapour from the air, and because it separates from the gasoline with which it has to be mixed. Its cost is at present much higher than that of petrol, and it is likely to be of most value as a fuel in tropical countries where vegetation is profuse.

In 1938, about 30,000 tonnes, i.e. about 0·5 per cent of the motor spirit consumed in Great Britain, was alcohol, in various blends with petrol. In France in 1955 nearly 100,000 tonnes of alcohol was consumed as fuel or for power, but the amount has varied very much from year to year.

Manufactured Gaseous Fuels

Coal gas began to be used in the early nineteenth century, not as a fuel proper, but as an illuminant for use in batswing and fishtail burners without incandescent mantles. It was not until the second half of the century that the advantages of gaseous fuels began to be generally realized. They are smokeless, ashless, and can be made sulphur-free; they have all the virtues of being "on tap"; and they can be used in very efficient appliances, internal combustion engines as well as furnaces. It is not surprising that, along with electricity, gas has revolutionized cooking methods, domestic hot-water supply, and domestic heating.

About 25 per cent of the calorific value of good gas coal was recovered in the form of gas, 50 per cent in the form of coke, and 7 per cent as tar and other by-products. From 5 to 10 per cent of the revenue of gas works arose from the sale of by-products other than coke but quite apart from this, the carbonization of bituminous coal was an economical way of using its calorific value provided there was a satisfactory outlet for the coke.

In the 1960s production of town gas from coal was superseded partially by its production from oil and then totally by natural (North Sea) gas.

Producer gas. If the right amount of air is blown through red-hot coke at about 1000°C, the effluent gases are chiefly carbon monoxide and nitrogen. This mixture is called producer gas, and the plant for making it is called a gas producer (Fig. 14).

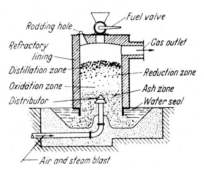

FIG. 14. Gas producer

In all large coke-burning producers a proportion of steam is introduced with the air which is blown through the red-hot coke. The products then contain a little water gas (see below), and there are considerable advantages in this system. The thermal efficiency of the process is raised; the calorific value of the gases in increased by the presence of hydrogen; and troubles due to the formation of clinker are reduced, because of the heat absorbed by the water-gas reaction.

Though its calorific value is relatively low (about $5 \cdot 6$ MJm^{-3}) producer gas has the great advantage over water gas that it is made in a continuous rather than an intermittent process. It can be made from almost all solid fuels, from waste such as cotton seed, and even from colliery refuse containing 50 per cent of mineral matter. Producer gas which is free from nitrogen, and whose calorific value is 9·3 MJm^{-3}, can be made by blowing steam and oxygen instead of steam and air through the coke.

Industrial producer gas is often made from coal, in which case it is enriched by the coal distillates; but the tarry matter and ash which such gas contains must be removed before the gas is suitable for use in engines and other specialized equipment. If tarry producer gas from coal passes through cool pipes, tar condenses in the pipes and ultimately chokes them. The simplest way to remove this tar is by burning it in the pipes; but "tar burning" is most objectionable because dense clouds of smoke are produced.

The original use of producer gas was in metallurgical furnaces of the regenerative type, which were invented in 1861; Another major use was for heating the retorts, chambers, or coke ovens in gas works. It can also be used in gas engines and in emergencies for ships, motors, tractors, and road vehicles.

Blast-furnace gas is the residual gas which passes out of the top of a blast furnace. In a blast furnace, the ore is reduced from oxide to metal by freshly made producer gas, only a part of which is used in the chemical reaction. Blast-furnace gas is therefore similar in constitution to producer gas, but it contains a higher proportion of carbon dioxide. It retains about 60 per cent of the heat of the original coke, and its calorific value is 70–80 per cent of that of producer gas, so it should not be wasted, though in older plants most of the blast-furnace gas was either allowed to burn at the top of the furnaces or led through flues to the atmosphere.

It can be burnt to provide the supplementary heat and power required by the blast furnace, but less than half of it can be effectively used in this way. To provide power, it can be cleaned of solids and used in gas engines, or it can be burnt to raise steam. Blast-furnace gas can be used in regenerative furnaces, and it is finding increasing use in the heating of coke ovens.

Sewage-sludge gas. At a number of modern sewage works, as the solid sediments are changed by bacterial action into an inoffensive form, the gases evolved are collected for use as fuel. The solids are kept at 27°C for 3–4 weeks in sludge-digestion tanks equipped with gas holders. The gas, which is approximately 70 per cent methane, is burnt in internal combustion engines to generate electricity for use about the works. It is estimated that 135×10^8 MJ per year would be available if all sewage in Britain could be so treated.

Water gas. If steam is blown through red-hot coke at about 1100°C, a simple chemical reaction occurs:

$$C + H_2O = CO + H_2$$

and two combustible gases, carbon monoxide and hydrogen, are formed. A small amount is also made of the non-combustible gas carbon dioxide:

$$C + 2H_2O = CO_2 + 2H_2.$$

The emerging mixture of gases is called water gas. More technically it is called blue water gas, to distinguish it from when it has petroleum additives (see below) and is then called carburetted water gas.

Water gas used to be produced in large quantities at British gas works.

Historical note

Town gas was the name given in Britain to the gas supplied by the Area Gas Boards, and was derived from the following sources:

(1) the Boards' own works which comprised carbonizing, water gas and oil gasification plants;
(2) the purchase of surplus gas from coke ovens owned by the National Coal Board, the iron and steel industry and certain independent companies;
(3) the purchase of tail gases from oil refineries;
(4) the purchase of liquefied petroleum gases; and
(5) small supplies of natural gas and of methane drained from mines.

The following table (Table 10) is partly from the Gas Council Annual Report 1958–9:

TABLE 10

Year	Gas made (million therms)					Gas bought		Total gas available
	Coal gas	Water gas	Oil gas	Other gas	Total	From coke ovens	From oil refineries etc.	
1953–4	2010	358	2	26	2396	348	5	2749
1956–7	2070	278	5	29	2382	457	10	2849
1958–9	1848	396	42	24	2310	465	65	2840
1958	Natural gas sold in U.S.A. for comparison							115,000

In the table, the term water gas covers blue gas as well as carburetted water gas (see previous section)—the quantity made depended not only on the demand for gas but also on coke sales. Oil gas includes production of "Segas" by the South Eastern Gas Board and "Onia-Gegi" introduced from France. "Other gas" includes refinery gas, methane and butane.

The Gas Act contained three requirements for town gas: (1) its calorific value had to be maintained at the declared value, (2) it had to be free from hydrogen sulphide, and (3) the gas pressure had to be not less than 2 in. water gauge in the main. The declared calorific values were mostly within the range 450 to 520 B.t.u./ft^3 and about half the gas was supplied at 500 B.t.u./ft^3 (500 B.t.u./ft^3 ≈ 19 MJ m^{-3}).

In addition to these three statutory requirements the organic sulphur content of town gas was kept as low as possible by the selection of coals and by use of sulphur removal plants.

For purposes of identifying gas quality and combustion characteristics the Gas Boards (now British Gas) adopted a gas grouping system based on the Wobbe Number. This is given by the expression calorific value divided by the square root of specific gravity The gas groups are given in Table 11.

TABLE 11. *Gas groups*

Gas group	Wobbe number	
	Mean	Range
G.4	730	700–760
G.5	670	641–700
G.6	615	591–640
G.7	560	530–590

Town gas production resulted in fuels of different Wobbe number, although 50 per cent was G.4 and over 98 per cent was covered by groups G.4, 5, 6. Nowadays North Sea gas has a consistent Wobbe number.

The flame characteristics of gases with different Wobbe numbers necessitate an appropriate design of burner. This range of town gas supplies was thus matched by burners marked with the appropriate gas group. Conversion to North Sea gas involved large-scale change of burners throughout the country—anyone wishing to convert nowadays to, for example, butane or propane (gases with very high Wobbe numbers) must also be prepared to install new burners.

CHAPTER 6

Combustion and Power Generation

THE preceding chapters have discussed the various primary fuels and some of their derivatives (e.g. coke). Usage patterns for these fuels have varied considerably over the years and it is relevant to first summarize in this chapter these variations as a prelude to a more direct consideration of their combustion in boilers (Chapter 7) and industrial furnaces (Chapter 8) and in their application in atmospheric pollution; and then to consider power production as it affects pollution levels.

Prior to 1800 only 12 million tonnes of coal were burnt per year. An increasing population together with an increasing degree of industrialization saw this figure rise to 170 million tonnes in 1900. At that time, other fuels held a negligible share of the market. Coal production reached a peak of 217 million tonnes in 1956 but this does not reflect the total combustion-based processes in the United Kingdom since the increasing use of oil (from 3 million tonnes in 1920 to a peak of 160 million tonnes in 1972) distorts the comparison.

Attempts to display these figures in an easy way for true comparisons vary tremendously. It was established, for instance, that in 1949 each individual in the United States had energy equivalent to twenty-seven servants working for him; the corresponding number in Great Britain, which came next in the utilization of fuel and energy, was thirteen to fourteen servants for each individual.

Nowadays, conversion of totals of oil, gas, etc., to their "coal equivalent" values is often used (see Table 12). In these terms it is simple to assess trends in various forms of energy. In the U.K. both the annual and the year-to-year variations can be easily compared (see Fig. 15).

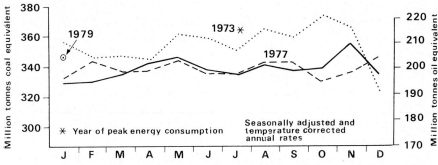

FIG. 15. Total United Kingdom energy consumption (after Department of Energy)

Combustion and Power Generation

TABLE 12. *Coal equivalents of various fuels*

Fuel		Coal equivalent (tonnes)
Type	Amount	
Petroleum oil	1·0 tonne	1.7
Natural gas	8076 kWh = 276 therms	1·0
Nuclear and hydroelectricity		Amount of coal to produce same amount of electricity at modern coal fired power station efficiencies

Since 1954 United Nations Statistical Papers, Series J, have summarized world energy supplies. Table 13 shows an example of how the world consumption of energy has changed since 1929, all values expressed in million tonnes of coal equivalent.

TABLE 13. *World consumption of energy (million tonnes of coal equivalent)*

	1929	1937	1949	1958	1975
Solid fuels	1367	1361	1482	1983	2500
Liquid fuels	255	328	557	1136	4000
Natural and imported gas	76	115	239	505	1700
Hydro and imported electricity	14	22	38	75	540
Uranium	—	—	—	—	570

The figures in Table 13 include energy used for heating and other purposes besides power. In the present chapter we are concerned with the production of controlled mechanical power (cf. the use of boilers and furnaces primarily for the production of heat). The most important prime movers may be represented by reciprocating steam engines (though these are declining in importance); turbines of various kinds, and internal combustion engines.

The Reciprocating Steam Engine (see Fig. 16) is a cylinder in which a piston, connected by a rod to a crankshaft on which is mounted a wheel or flywheel, can move backwards and forwards. Steam under pressure from a boiler is admitted to the back of the cylinder and the piston is pushed forwards, rotating the flywheel about one-third of a revolution. The steam is then shut off by a valve and the piston continues to be pushed forwards by the expansion of the steam in the cylinder. As it expands, this steam cools and begins to condense into droplets of water. At the beginning of the backward stroke a second valve opens to admit steam to the front of the cylinder, and the piston is pushed backwards. The condensing steam at the back of the cylinder escapes into a "condenser" where it is completely converted to water, the vacuum created helping to draw back the piston. The condensed water is pumped into the boiler against the pressure of the steam by the boiler feed-pump. The valve mechanism is very robust, a single slide

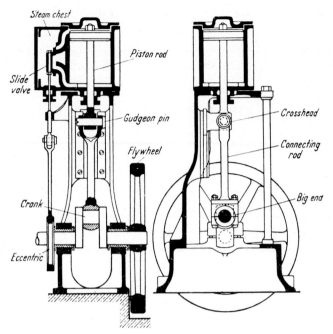

Fig. 16. The reciprocating steam engine

performing the operations of connecting each end of the cylinder alternately to the steam pipe and the condenser with suitable intervals for the expansion of the steam in the cylinder.

The steam engine is a "heat engine", deriving its energy from the difference in heat content between hot steam and boiling water. When the temperature difference is 100°C a maximum of about 30 per cent of the heat given to the steam is recoverable as power. The fraction of the fuel's energy which can be recovered from a reciprocating steam engine is seldom more than 15 per cent, and efficiencies of 3–12 per cent are more usual. There was no room in locomotive engines for condensers, and their over-all thermal efficiencies were consequently low, 3–6 per cent. In most steam engines about 25 per cent of the energy was lost in the boiler, and the rest in the engine.

The Turbine is a more efficient machine for using steam, where 100 to 100,000 kW are required. It is able to convert into useful work the energy in the steam as it expands right down to the pressure in the condenser. The turbine is more compact than the ordinary steam engine, and rotates more smoothly.

In operation, jets of steam are directed between curved blades on the rim of the rotor. The steam is made to change its direction of motion and also to expand, doing work on the rotor. A two-pole 50-cycle turbo-alternator rotates at 3000 rev/min, a four-pole alternator at 1500; in America, where 60-cycle machines are common, the corresponding speeds are 3600 and 1800 rev/min. Until recently the usual efficiencies were from 12 to 30 per cent, but efficiencies of over 40 per cent are now possible. The efficiency of a heat engine depends on the ratio of the highest to the lowest temperature (in degrees above absolute zero) reached by the working substance.

The Gas Turbine has been developed as a power unit since 1940, although the principles controlling it have been known for nearly 200 years. Although gases and powdered solids are possible fuels, liquid fuels are the most convenient, and they are burnt in compressed air in such a way that the gaseous products of combustion become very hot. If these were retained in the combustion chamber the pressure would increase even more, but the gases are made to escape between turbine blades, expanding and cooling as they do so. The gas turbine is relatively light, because combustion takes place near the rotors. It is efficient only if the gases are allowed to reach very high temperatures, between 600 and 850°C, and materials have not been available until comparatively recently for making a rotor which will operate satisfactorily within this range of temperatures. A 4 MW gas turbine was built in 1939 in Switzerland for use as a stand-by electric generator.

Since 1944 there have been great developments in small gas turbines for aircraft, naval vessels, railway locomotives and other uses. These engines can generate great power in proportion to their size and are particularly suited to "jet" propulsion. They burn kerosene, which is both cheaper and safer than gasoline, or a mixture of the two. The use of jet engines is widespread for aircraft.

Internal-combustion engine

The gas turbine is an internal-combustion engine, since the fuel is burnt within the turbine space, but the phrase is generally restricted to piston engines in which the fuel is burnt inside the cylinder. Any gaseous fuel, such as producer gas, sewage-sludge gas, or the vapour of petrol or oil, can be made to burn very rapidly, in fact explosively. This is done by mixing fuel and air in the right proportion, raising the temperature above a minimum which depends on the fuel, and by applying a flame or spark if the temperature is insufficient for spontaneous ignition. The same is true for liquid and solid fuels when these are broken up into droplets or particles less than about 25 μm in diameter.

As a result of such rapid combustion it is possible to make effective reciprocating engines in which the gaseous products of combustion are the working substance instead of steam, combustion taking place inside the cylinder instead of under a separate boiler. Internal-combustion engines are classified according to the method of igniting each charge of air and fuel mixture (spark ignition or compression ignition) and according to the cycle of operations undergone by the mechanism (four stroke or two stroke).

The four-stroke engine works with the help of two valves to each cylinder. On the charging stroke, the piston acts as a pump, sucking the mixture of air and fuel into the cylinder through the inlet valve. On the return stroke it compresses and adiabatically warms the mixture, which is then ignited. A rapid rise in temperature and pressure results and the working stroke begins, during which the mixture of gases expands and partly cools. On the return from this stroke the exhaust valve is opened, and the piston drives out the spent gases. The exhaust valve closes and the cycle is repeated.

In the two-stroke engine (see Fig. 17) the under-side of the piston is used to compress the mixture of air and fuel during the down stroke. Towards the end of the down stroke the piston uncovers ports in the cylinder wall and the fresh mixture enters the cylinder head, driving out the spent gases. A proportion of spent gases inevitably remains, but though the explosions are less powerful they occur twice as often as in the four-stroke engine, and the final efficiency is about the same.

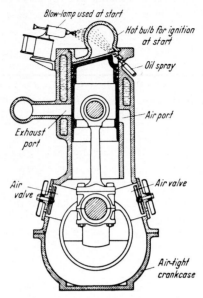

FIG. 17. A simple internal-combustion engine. Two stroke, compression ignition

Gas engines are rather less wasteful of fuel than steam engines. Their efficiency is 25–30 per cent. Small motors driven by petrol or oil are about as efficient as gas engines, their efficiency being 18–25 per cent. Larger engines of the compression-ignition (diesel) type have efficiencies ranging from 25–33 per cent. They are at present the most efficient machines for converting the energy of fuel to mechanical energy, though they are not always the most suitable because of fuel and upkeep costs.

Atmospheric Pollution from Engines

It is evident that large engines of all types are usually more efficient than small ones. As a matter of world and national fuel economy, therefore, it is important to develop large power plants. This is also a good policy when atmospheric pollution is considered, because in a large plant it is worthwhile to take very strict precautions against emitting pollution.

There is a general tendency for small steam engines served by coal-fired boilers to be replaced by other systems, including electric motors dependent on power from big coal-burning electric generating stations. Except for the effects from cooling towers which are discussed below, the problem of pollution from steam engines is no different from the problem of pollution from boilers. At one time the nuisance was proportionately greatest from railway engines, in which nearly 10 million tonnes out of a total of about 200 million tonnes of coal were burnt in 1960 in Britain. This was 5 per cent of the total coal consumption, but the corresponding figure in 1949 was 7·5 per cent (see Table 14). It is difficult to estimate the total weight of smoke that was emitted from this 10 million tonnes of coal but probably the amount was about 0·3 million tonnes. The emission of sulphur dioxide would also be about 0·3 million tonnes. As to grit and dust, an estimate of 0·1 million tonnes would not have been too high since forced draught was used in railway engines. The nuisance of all three types of pollution was aggravated by the unavoidable shortness of the smoke stack, though it is true that much of the emission occured in engine sheds or in open country (see also Chapter 10).

TABLE 14. *Quantities of coal used in the U.K. in 1938, 1956 and alternate years from 1960 to 1974 in grates or furnaces that may cause smoke emissions* (million tonnes)

Year	Domestic excluding miners'	Miners' coal	Railways	Power stations	Industrial* miscellaneous and collieries
1938	47·3	4·7	13·6	15·3	61·1
1956	31·1	5·4	12·3	46·3	63·6
1960	29·6	5·1	9·1	53·5	48·3
1962	27·9	4·8	6·2	61·4	43·9
1964	23·4	4·4	3·9	68·5	40·5
1966	21·2	3·9	1·7	69·0	36·6
1968	18·4	3·3	0·2	75·6	29·7
1970	15·4	2·7	0·1	77·2	25·6
1972	10·6	2·2	0·1	66·7	15·9
1974	9·8	2·1	0·1	66·0	14·5

* Excluding coke ovens, gas-supply industry and plants for making solid smokeless fuels.

Atmospheric pollution from internal-combustion engines consists of unburnt hydrocarbons, oxides of nitrogen and carbon and, sometimes from diesel engines, carcinogenic aromatics. Emissions depend upon whether the vehicle is accelerating, cruising or decelerating. Emitted near street level, or in garages or tunnels, exhaust gases can be dangerous. Concentrations of 1 per cent of carbon monoxide in air are rapidly fatal, and deaths from poisoning by carbon monoxide have been frequent. Fatal concentrations are likely to occur only in confined spaces, and can be prevented by adequate ventilation. All forms of pollution from traffic, including the formation of nitrogen oxides and complex organic compounds such as the irritant peroxyacetyl nitrate (PAN), are discussed in detail in Chapter 10.

Well-maintained diesel engines emit on average less hydrocarbons and similar amounts of nitrogen oxides compared to petrol engines. Poorly maintained engines, or engines working beyond their design capacity, can produce smoke which contains potentially harmful organics such as pyrene, fluorene, coronene, anthracene and 3–4 benzpyrene (a known carcinogen).

Local emissions are considerably less with many of the experimental internal- and external-combustion engines and especially with electrical systems. Control devices can be implemented and much has been done recently by legislation in the United States towards minimizing pollutants from internal-combustion engines.

Emissions from (jet) aircraft engines are largely hydrocarbons whilst idling and nitrogen oxides whilst cruising. Supersonic transport, SST (e.g. Concorde), fly above the tropopause, in the stratosphere. Pollutants emitted at this height are effectively trapped there. However, at this height exists an ozone layer which protects life on Earth by absorbing incoming, deadly ultra-violet radiation. Nitrogen oxides, together with the small amount of water vapour emitted by SST's (which is disproportionately important due to the almost complete absence of naturally occurring water vapour in the stratosphere), react with the ozone and have been accused of destroying this protective layer. Our knowledge of stratospheric chemistry is still extremely limited and it is not yet known whether or not there exists a self-regulatory mechanism that will restore ozone concentrations.*

Cooling towers

For thermodynamic efficiency, all steam engines require condensers, where exhaust steam is quickly changed to water, thus creating a vacuum which heeps the motion of the piston or turbine. It is also an advantage to re-use the condenser water in the boiler for making steam, since this is "distilled" water that is free from dissolved solids. The condensers are kept cool by an independent water supply circulating through numerous small tubes across the space in which the steam is condensing. This condenser water need not be pure, and it is drawn from town mains, rivers, or canals; but it is wanted in large quantities, about 600 tonnes for each tonne of coal burnt in the boilers. Where mains water is used or where the natural supply is limited, the condenser water is circulated through cooling towers, where it is cooled by air and its own latent heat of evaporation, so that it can be used again. Cooling towers, particularly the large ones used by power stations, tend to produce a fine mist which may fall outside the precincts. In accordance with the definition at the beginning of Chapter 1, this is atmospheric pollution, and therefore calls for some further consideration.

Although forced-draught cooling towers are common in the U.S.A., natural draught is generally employed in Britain. In a wet, natural draught cooling tower, the incoming water is distributed by sprays or other devices on to a wooden packing over whose many surfaces it flows down to a sump, from which it is pumped for recirculation through the condensers. At the packing, heat is transferred from the water to the air, and the warmed air rises, drawing fresh air through the inlets at the base of the tower. A natural draught is thus produced by the difference in density between the cold air outside and the warm air within the whole height of the tower; the air moves up the tower with velocities usually of $1-3$ m s^{-1}. The temperature rise of the air and the fall in water temperature are quite small, perhaps 15-20°C, but the volumes of air and water are so great that a very large quantity of heat is dissipated.

---

* The release at ground level of chlorofluorocarbons from some aerosol cans has similarly been Once again we do not have sufficient information on the chemical kinetics of such compounds either in the troposphere or in the stratosphere to calculate their possible climatological effects.

The air becomes saturated with water vapour as it rises through the packing, and water droplets also are entrained in the air stream. In a typical modern cooling tower, perhaps a hundredth of the water circulated is lost by evaporation, and a thousandth by entrainment. Some of the entrained water falls back at once to the packing, but the smaller drops are lifted right out of the tower. In the open these may at first grow in size by condensation from the saturated air around them, now beginning to cool, but they soon fall out of the saturated air, with velocities comparable with the upward air velocity in the tower. If the normal atmosphere is unsaturated the droplets will partly evaporate before reaching the ground.

The nuisance from cooling towers is now thought to be due almost entirely to the entrained droplets and hardly at all to the water vapour in the warm air of the tower. If there were no droplets already present, some of the water vapour would condense on reaching the open, but the droplets then formed would be extremely fine, and would have time to mix with fresh unsaturated air and evaporate before they could reach the ground.

The mist from cooling towers is usually worst in cold weather for three reasons: (1) the air flow in the towers tends to increase when the incoming air is cold, (2) the demand for electricity is high and (3) the evaporating power of ordinary air, even if the relative humidity is low, is much reduced in cold weather. It is questionable, however, whether for efficient generation of electricity the cooling-tower water needs to be cooled by the same amount in cold as in warm weather, and one way of reducing the mist in cold weather might be to throttle the air flow into the base of the tower, or lead a proportion of the water through a by-pass instead of through the tower.

The most obvious way of eliminating the nuisance is to reduce the load on each tower by providing extra towers, but less costly methods are preferable. Assuming that the trouble is due to entrainment rather than evaporation it may be attacked in at least three other ways: (1) by ensuring that the air flows uniformly through all parts of the packing and that there are no places where the flow is excessively fast, (2) by distributing the water over the packing without splashing, and (3) by trapping entrained water with wooden baffles or other "eliminators". The third method has already been found effective in America. As a result of research now in hand by the British Central Electricity Generating Board and other bodies there is a good hope not only that mist from cooling towers will be prevented but that their general efficiency will be improved.

As a fourth alternative, a heat exchanger can be used in which the water is retained inside the heat exchanger and heat is lost to tha rising air current across a solid boundary by condution, with the air rising naturally or with the help of blowers. The design of the heat exchanger is critical and still the subject of much intensive research.

Electricity

Since the earliest days of engines one of the main problems has been to bring their power to the place where it is wanted. The locomotive, the crane, the lathe, and the riveting machine, for example, call for widely differing methods of transmitting power. In a factory where many machines are liable to be in use it is often an advantage to make a single engine drive a number of machines. This was generally done by using the engine to drive a line of "shafting", which by the use of fixed and loose pulleys could be geared independently to each machine.

Although there are many processes and situations where such methods are still the best available, recently there has been a strong tendency for mechanical to be superseded by electrical transmission of power. Electric power can be transmitted long distances along wires. Electric motors are relatively cheap and their efficiency is high, from about 70 per cent for a 200 W motor to 90 per cent for a 100 kW motor. It is often economical, instead of using a line of shafting, to drive each machine with a separate electric motor. Also a greater variety of machine tools is possible and their layout in the factory is more flexible.

Many of the electrical properties of matter were known in early times, but it was not until Faraday studied the relation between electricity and magnetism that the possibility of making dynamos and motors was realized. Electricity then began to come into its own as a highly convenient method of transmitting rotary power. At the power station a dynamo is rotated by any suitable means. Two, three or four wires connect the dynamo to a number of electric motors and each of these rotates, when required, on the closing of a switch.

Dynamos are rotated by many sorts of engine, including each of the major types described earlier in this chapter. The biggest output of electric power is from dynamos driven by steam turbines, but the demand for electricity seems insatiable and every economical means of producing it is called into service (mainly from combustion of fossil fuels). In several countries much more fuel oil than formerly is being used for steam raising at power stations, and nuclear energy is also coming into use. The growth of steam turbines and water power in Great Britain is shown in Table 15. The annual output of electric power has been doubling every ten years since 1920 until the peak year of 1973 after which electricity generation has levelled off. This has been in line with the rate of increase in the world as a whole—over 10 per cent per year.

TABLE 15. *Electricity generated in Britain*
million kWh/year

	1920	1930	1940	1950	1959	1967	1974	1977
Steam	4200	10,400	27,800	53,300	102,900	176,900	232,400	242,500
Hydroelectric	12	320	800	1035	2175	16,100	13,500	12,500

In mountainous countries, hydroelectric power stations are relatively more important than in Britain. They are often associated with an industry which requires much cheap power, e.g. aluminium, magnesium etc.

Unlike steam, water does not contract appreciably under pressure, and it is more liable than steam to flow with a turbulent motion and thus to dissipate its energy. For these and similar reasons, water turbines are different in detail from steam turbines, though they are similar in general principle. The capital costs of hydroelectric stations are two to three times those of thermal stations, but their operating costs and rate of depreciation are remarkably low. Because of the great ease with which water turbines can be started and stopped, the tendency is for hydroelectric stations to be used for peak load, instead of being kept running day and night (base load) like some of the most efficient thermal stations.

In England and Wales, a 360 MW "pumped storage" scheme is at present in operation near Blaenau Ffestiniog, and several similar schemes are in plan form. Unwanted grid electricity is used to elevate water from the lower reservoir to the upper. Release of this water at times of peak power demand permits conversion of this potential energy back to electricity by use of underground turbines (see Fig. 18).

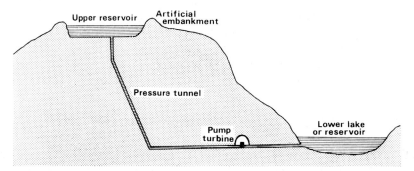

FIG. 18. Pumped storage scheme for hydroelectric power (e.g. Blaenau Ffestiniog)

Power generation by waves and tides is currently being developed and La Rance station in France (a tidal estuary scheme) is at present supplying 550,000 MWh per year and a tidal power scheme on the Severn Estuary holds the promise of a large energy production for Britain in the near future.

Uses of electricity

When Michael Faraday spun a metal disc in the field of a magnet and produced an electric current, he changed the direction of civilization. It would now be very hard to imagine what our industries would look like, or what articles would be in the shops, and their price, in the absence of electric power stations. For some purposes there is literally no substitute for electricity, and for many others the substitutes are clumsy, dirty and expensive. The main uses of electric power from generating stations, excluding scientific research, can be divided into six classes.

Electronic uses include communications by telegraph, telephone and radio, computers automation and servo-systems, television and radio. Though power consumption is small in all these fields their usefulness is great.

Lighting nowadays is almost entirely by tungsten filament lamps, discharge tubes and fluorescent tubes. The conversion of electrical energy into light is many times more efficient, and cleaner and simpler, than any other way of making artificial light. Again, however, lighting is a relatively small user of electric power.

Electrolysis is one way, and in some important cases the only practical way, of decomposing chemical compounds into their constituent elements or into other substances which are rare in nature. It is used for making soda ash, caustic soda, chlorine, sodium hypochlorite, and for purifying copper and zinc. But for electrolytic methods, we should have virtually no aluminium, magnesium or titanium in pure metallic form.

Power can conveniently be distributed, over long distances, only by fluids and electricity. Once electric power has been "laid on" it can be obtained when required, in the works, factory or home, with the minimum of effort. So easy is its distribution and use that it can be used for traction in trains and buses, and it has even been used to replace the mechanical transmission system of ships' engines. The keenest competitor of the electric motor in general is the oil or petrol engine, and which is preferable depends on the place and the type of application. In the present epoch, oil and petrol engines are competitive whenever they have to be mobile, but electrical power units are often cheaper when they are to be used in fixed positions, especially of course in countries where plenty of hydroelectric power is available. What the position will be when the conversion of nuclear energy to electric and other power becomes acceptable, or when fossil oil fuels begin to become scarce, is now the subject of debate.

Specialized heating purposes where electricity is essential or advantageous are many. Electric furnaces are capable of temperatures as high as 3400°C without serious contamination of products, whereas combustion furnaces seldom exceed 1650°C. The heaviest users of electric furnaces are the steel industry (for casting steel and for producing ferrosilicon, ferrochromium, and other compounds needed in making high-grade steels), and the chemical industries for producing phosphorus, calcium carbide, carbon bisulphide and other materials. Smaller electrical appliances under this general heading include irons and soldering irons, grillers, toasters, specialized baking ovens, kettles, small boilers and hot air driers; many of these and other appliances have their counterpart which uses gas or some other fuel.

Space heating is the field where competition from other sources of heat is most effective. The losses of a heat engine have made it difficult to produce electric power from coal or other fuels without wasting nearly 70 per cent of the calorific value of the fuel. In the future we may hope, however, for high-pressure, high-temperature steam stations, which waste only 60 per cent of the fuel, or for fuel cells which generate electricity by chemical action and are even less wasteful of natural fuel. Electric power may also be converted into heat at over 100 per cent efficiency if it drives a heat engine in reverse, but this entails rather elaborate equipment.

On the other hand there are many arguments and applications in favour of electricity. Unlike combustion heaters, except oil and gas, it can be switched on and off. It can indeed be made very cheaply from water power, or more expensively (at present) from tidal, wind or solar power. When made from coal it uses low-grade small coal which is mined in great quantity but which cannot easily be disposed of except to power stations.

One important argument justifying electrical heating for normal purposes, particularly for overnight heating of buildings, arises, oddly enough, from the main drawback of electricity. (See Fig. 19, examples of the fluctuating demand for electricity.) It cannot be stored in useful amounts, except by such expedients as using it to drive water uphill into reservoirs, which may then serve hydrostations. Neither can a giant coal-fired power station, whose boilers may take 6 hr from cold to raise full steam pressure, be fully shut down overnight. Hence many generating stations must be "banked" overnight.

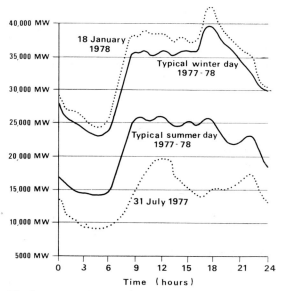

FIG. 19. The fluctuating demand for round-the-clok electricity (CEGB). Compare the 13,000 MW rise in winter between 5 h and 9 h with the 11 GW (11,000 MW) fall between 0 h and 9 h in off-peak demand in Fig 20.

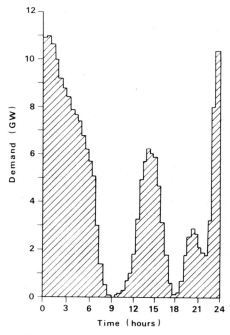

FIG. 20. Total off-peak demand on a typical winter day 1972–73 (CEGB)

when the demand is least. Any "off-peak" use of electricity helps to reduce the overall cost of generation. As a result, both industrial and domestic consumers employ time switches to operate storage heaters etc. especially during the period 1.30 to 8.30 a.m. (see Fig. 20) where peaks of 11,000 MW are supplied. (A large increase from only 300 MW supplied in 1958.)

One of the most recent phenomena in variability of electricity demands is the television—programme "ends" result in huge surges in power demands (see Fig. 21 where the end of the 1972 Miss World contest on BBC 1 resulted in an increase from 30·5 to almost 32·0 GW).

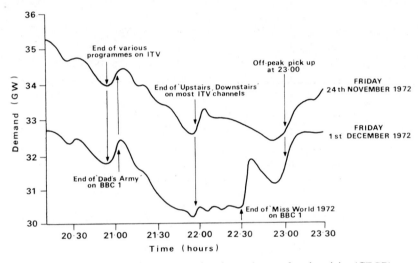

FIG. 21. Effect of television on evening demand curve for electricity (CEGB)

Conclusion

The case for electricity in any context, particularly in this book, cannot be completed without a consideration of the amounts of dirt and pollution which are produced in the generation and use of electricity, and in the production of alternative forms of power and heat. Here, purely economic discussions tend to go astray unless the cost of pollution is included; whereas it is mainly excluded in the more or less open competition of the fuel market. However, in the present book, discussion of atmospheric pollution in relation to its cost and prevention must be deferred (until Chapter 15) after our survey of the commonest boilers, furnaces and heat services.

CHAPTER 7

Industrial Boilers

GENERATION of steam by the heating of water in a boiler necessitates a heat source. This is usually provided by combustion of fossil fuel (alternatively, for example, nuclear fuel could be used). Heat generation by combustion is used in many industries, not only for steam raising but in ore refining, pottery manufacture, etc. These latter uses are discussed in Chapter 8 which describes furnaces and kilns.

There are two basic boiler types: shell and water-tube. The former are relatively inexpensive with internal fire boxes; the latter more useful for providing high efficiencies and maximum pressures. Shell boilers can be further subdivided into vertical, horizontal (e.g. Lancashire, Economic) and waste-heat boilers. Water-tube boilers may either be subcritical (in which the water is recirculated) or supercritical (once through).

Coal-fired boilers

A later section deals with hot-water boilers for industrial space heating. We are concerned here with industrial boilers for steam raising. Oil-fired boilers, having been dealt with in Chapter 4, will receive only a passing reference, and most attention will be paid to the burning of coal.

Table 16 provides information about some of the commoner types of industrial boiler now in use. The choice of a boiler depends on the amount, temperature and pressure of steam required; the variations in the steam demand, including opportunities for shutting down and cleaning; the space available; the alternative fuels; the quality of the water; and general economic considerations. An engineer making such a choice would calculate the size and type of boiler from information similar to that of Table 16, assuming conditions of easy steaming which have been assumed in the table. Under heavy steaming the output may be increased by 30 per cent or more for short periods according to the type of boiler. He would then review the characteristics of each boiler in greater detail, bearing in mind the possibilities of alternative fuels such as oil or natural gas.

TABLE 16. *Characteristics of boilers*

Boiler and dimensions	Grate area (m²)	Easy steaming output (kg s⁻¹)	Steam pressure (N m⁻² ×10⁵)	kg steam produced per kg coal	Thermal efficiency (%)
Vertical:					
1 m dia. by 3 m high to 2 m by 6 m	0·2 to 4·7	0·1 to 0·8	5·5 to 10·3	5 to 8	40 to 65
Lancashire:					
2 m dia. by 8 m long to 3 m by 10 m	2·0 to 4·7	0·4 to 1·5	6·9 to 17·2	6 to 9	50 to 70
Economic:					
1½ m dia. by 2½ m long to 4 m by 6 m	0·9 to 13·9	0·09 to 3·8	6·9 to 17·2	7 to 10	60 to 80
Water-tube:					
setting 1 m wide by 1 m long by 3 m high to 8 m by 11 m by 27 m	0·6 to 92·9	0·04 to 126·0	17·2 to 96·5	7 to 11	60 to 90+

Note: The details of performance given in the four right-hand columns are approximate.

Vertical boiler

In the simple cross-tube boiler the hot gases pass directly from the fire box into the uptake, the lower part of which, like the fire box, is encircled by the water space. Water also fills the cross-tubes, one or more in number, which are slightly inclined to the horizontal to promote circulation. For the removal of scale resulting from the deposition of solids from the water each tube requires a separate cleaning door in the outer shell of the water space. The main advantages of these boilers are their requirement of small floor space and their general simplicity: they are often used as crane boilers.

For the maximum proportion of heat to be transferred from the fuel to the water, it is important that (1) a large surface of the water space should be exposed to direct radiant heat from the burning fuel, and (2) as much heat as possible should then be transferred from the flue gases to the water.

An improved design of vertical boiler is the fire-tube type such as the Cochran illustrated in Fig. 22 and this accomplishes the second purpose by the passage of the hot gases along fire tubes which cross the water space horizontally. On opening the front of the smoke box, the insides of the fire tubes are easily accessible for cleaning. The Cochran boiler is reasonably efficient and because of its robust design maintenance costs are usually low. It is very useful for providing temporary supplies, taking up little floor space, but unfortunately cannot be adapted for oil firing.

Boilers used as crane boilers are usually "fed" by injectors but any boiler in continuous service requires a feed-pump to force water into the boiler against the pressure of the steam. Most recent installations have float-controlled feed pumps to maintain the water level within prescribed limits.

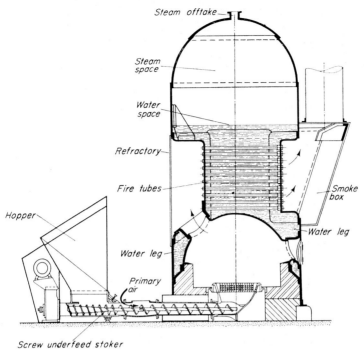

FIG. 22. Cochran boiler

Lancashire boiler

In Great Britain this is the commonest form of industrial boiler, though it is quite uncommon in most other countries. It will tolerate unskilful handling, and burn most ordinary fuels, as well as sawdust and other wastes; although designed for coal firing, it cannot be easily converted to oil firing. It may have a useful working life of over 40 years—much longer than the Economic boiler. Its chief disadvantage is its great size, which is increased by the setting and flues, with the superheater and economizers which are necessary if high thermal efficiency is to be attained.

The Lancashire boiler is of such dimensions that, during overhaul or cleaning, a man can reach any part. It has two independent furnace tubes, about 1 m in diameter, placed side by side within a horizontal boiler shell, of average size 2·5 m diameter by 8·5 m long. As Fig. 23 shows, the first 2 m of each furnace tube is divided by a horizontal grate into a combustion zone above and an ash pit below. At the inner end of the grate, the lower two-thirds of the furnace tube is filled by a firebrick wall: the "bridge". Normally the furnace door is kept closed and the entrance to the ash pit open. The primary air supply is admitted through the ash pit, and its amount is controlled by dampers in the side flues or a single damper at the base of the chimney. Flames and hot gases pass along the furnace tubes, and the gases are made to return under the bottom and then along the sides of the boiler, to give up as much as possible of their heat before escaping up the chimney. If superheaters are fitted in these flues they also remove some of the heat from the gases. The flues and setting are of brickwork, and they must remain airtight in spite

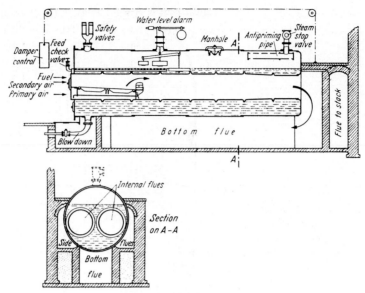

FIG. 23. Lancashire boiler

of the thermal expansion of the long boiler shell. The entry of air through leaks into the bottom- and side-flues is well known to be the greatest potential source of heat loss in a Lancashire boiler.

The gases escaping up the chimney are still hot, and a saving of fuel, often 5 per cent or more, is effected with economizers, which enable the flue gases to impart heat to the cold feed-water before it enters the boiler. Economizers can be considered as an extension to the boiler heating surface. A 6°C rise in feed water temperature saves approximately 1 per cent of the fuel. Air pre-heaters also are sometimes used.

The draught which draws the air through the furnace and flues is provided by the difference in density between the warm gases inside the chimney and the cold air outside. This natural draught may be supplemented by (1) an induced-draught fan placed within the stack, (2) forced-draught fans supplying air to the closed ash pits, (3) a combination of (1) and (2) so that the pressure in the combustion space above the firebed is zero, i.e. the same as in the air outside, this being known as "balanced draught" or by (4) steam jets inducing air into the ashpit.

Economic boiler

To use effectively the heat from the burning fuel, a Lancashire boiler must be long, and have an extensive setting with accommodation for economizers and, possibly, air pre-heaters. Consequently the complete boiler occupies a large total space. With the Economic boiler, illustrated in Fig. 24, the same thermal efficiencies are achieved without economizers or air pre-heaters, and with a great saving of space. It is perhaps the most frequently commissioned boiler type today but has a maximum working life of only 15 years.

The fuel is burnt in one or more furnace tubes similar to those of the Lancashire boiler, immersed in the same way in a horizontal boiler shell, though this is only about 5 m long. The hot gases finish burning in a combustion chamber water-cooled or lined with refractory bricks, at one end of the boiler, and return along horizontal tubes through the boiler shell; they may be returned through the boiler a second or even a third time, through other nests of tubes. Finally the gases enter a smoke box from which they escape up the chimney.

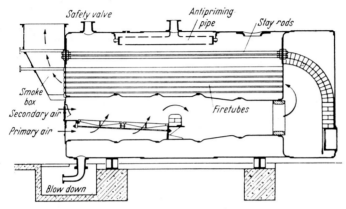

Fig. 24. Economic boiler

Apart from its combustion chamber of brick, the Economic boiler needs no brick setting, but it is less robust than the Lancashire boiler and has a smaller space for storage of water and steam. Soot is removed from the smoke tubes by brushing or by blowing steam or, better still, compressed air through them (soot blowing). Economic boilers need water as free as possible from dissolved solids; in most areas, the main water supply should be treated in an ion-exchange plant before use.

Thermal storage boiler

The steam demand in many factories varies widely owing to the intermittent operation of the different processes, giving rise to wide variation in boiler and process pressures; in addition, the rate of firing must be continually altered. Under these circumstances, loss of productive output and waste of fuel are likely unless special measures are taken.

The *thermal storage boiler* combines steam-raising and heat-accumulation in such a manner that its water volume is made to act in effect as a thermal flywheel, storing heat by controlled rise in water level during periods of low steam demand and releasing this heat by controlled drop in water level to meet the peaks. Its shell is enlarged in diameter compared with a normal boiler, to enable a safe rise and fall of water level of about 1 m to be obtained (see Fig. 25).

With the thermal storage boiler variations of 25–40 per cent above and below the average steam demand can be met without change either in process steam pressure or in firing rate.

Control is by the steam-pressure regulator A, and the water-feed regulator B. A modified water-level gauge, shown at C, is graduated and marked so as to give the boiler maintenance personnel specific instructions when to reduce the rate of firing as the water level approaches the top limit and when to increase the rate of firing as the water level approaches the bottom limit. If he fails to take action at the proper time, high or low water level over-controls come into operation and alarms are sounded.

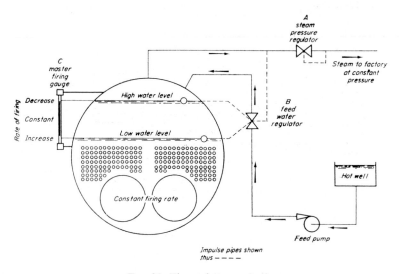

FIG. 25. Thermal storage boiler

Water-tube boilers

The Lancashire, Economic, and other so-called "shell" boilers have a large reserve of water in relation to steam space which gives capacity to meet variations in steam load with little alteration to the firing rates. This class of boiler is suitable for many heating and manufacturing purposes, particularly where steam is required at pressures of less than 14×10^5 N m^{-2}, e.g. "process" steam which is used simply as a carrier of heat, or steam used to drive a reciprocating engine. The size of shell boiler is limited by the weight of the shell and the cost per unit of performance.

Water-tube boilers can be used, given reasonable attention, in the place of shell boilers. Although they are costly, their cost increases more slowly, with output, than that of shell boilers; and where very large outputs of steam are required, at all pressures over 20×10^5 N m^{-2}, as in modern power stations, they are essential. They can be quite small or very large indeed, being assembled from bricks, tubes, drums, and other relatively small components, and they can be designed to burn almost any type of fuel. Both their water space and steam space are small, and as a result they can meet sudden fluctuations of up to 15 or 20 per cent in the demand for steam, but to achieve best results they require either continuous attention or automatic control. Further points are that their brick settings are complicated and must be frequently inspected, their feed water must contain the minimum of dissolved solids, and they require highly skilled attention.

Industrial water-tube boilers differ greatly in design, but there are two main classes: those with their water tubes expanded, as in Fig. 26, into sectional headers; or into three or more drums, as in Fig. 27. In the boiler of Fig. 26, water is pumped through the economizer into the saturated-steam and water drum, and circulates by convection through the bottom header, the tubes and the top header back to the drum, from which steam passes through the superheater to the steam line.

In the three-drum boiler illustrated in Fig. 27, feed-water from the economizer passes first through tubes embedded in the walls of the combustion chamber to the two water

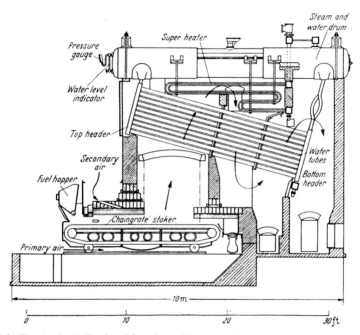

FIG. 26. Water-tube boiler (straight tubes with sectional-header with chain-grate stoker)

drums, and then through water tubes across the combustion space to the saturated-steam drum. Forced circulation is employed, for the sake of a saving in water space and a quicker response. Steam from the saturated-steam drum is driven through the superheater to the turbines. For highest thermal efficiency, it is now recognized that the maximum radiant heat must be transferred from the fuel to the water, and in large boilers the walls of the combustion chamber are water cooled both to raise efficiency and to prevent overheating.

The methods of firing, too, are various. The sectional-header boiler in Fig. 26 is shown with a chain-grate stoker; the grate moves slowly to the right, carrying coal from the hopper into the combustion zone, and tipping the residual ash over the far end into the ash pit. In the multi-drum boiler in Fig. 27 a mixture of pulverized coal and primary air is blown downwards into the top of the combustion chamber; secondary air is blown in through suitably placed tuyères. The heavier unburnt particles, mostly mineral matter, fall into the ash pit or are removed mechanically, but the lighter particles are swept in an upward curve by the air stream, and combustion continues until the first bank of water tubes is reached. Either the sectional-header or the multi-drum type of boiler may be designed for pulverized fuel or for the chain-grate, the retort stoker, or any other form of automatic stoker.

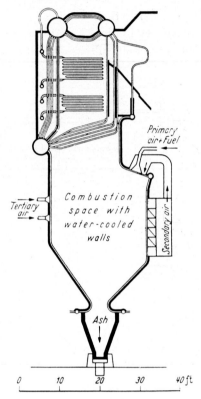

FIG. 27. P.F. multi-drum boiler

Industrial Hot Water Boilers

The heating of buildings by high-pressure hot water is rapidly supplanting steam heating. This is chiefly because the pipe circuitry is easier (steam installations require a system for collecting the condensed water); and water takes up less space than steam for a given quantity of heat carried.

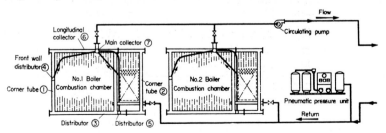

FIG. 28. Hot-water boiler

Both steam and hot-water systems are above the boiling-point of water, and pipes are too hot to touch. To reduce the temperatures of exposed surfaces below the safe limit of 82°C, calorifiers are used to transfer heat into subsidiary circulating water systems. Convectors may also be used in which air is blown past a coil of hot high-pressure pipes and into the space to be heated.

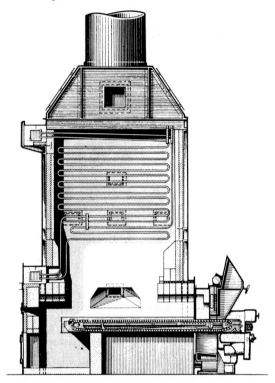

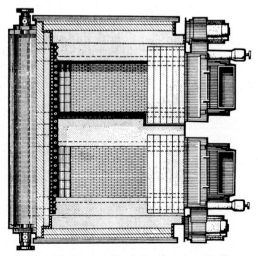

Fig. 29. Forced-circulation hot-water boiler

Hot-water boilers have the flow pipe connected below water level, and there can be a small steam space in which a pressure reserve is available to drive the water through the system. In the more usual method for large installations, the boiler is quite full of water and there is no steam space. A pressure vessel is kept filled to the desired pressure with air, or nitrogen to prevent corrosion. The water is forced round the circuit by a pump (see Fig. 28).

Various boilers have been designed specially for high-pressure hot water, with a view to maximum convenience in installation and minimum space requirements. The one illustrated in Fig. 29 incorporates a series of tube elements arranged in parallel (only two are shown in the figure). These are connected to top and bottom headers, the latter being connected to a main steam-and-water drum. Water is circulated through the headers, drum and system by pumps. Several boilers may be used in parallel, with a common drum.

The heating surfaces are enclosed in an insulated steel casing. There is suitable provision for cleaning the tube elements. This type of boiler is economical in (1) steel and (2) floor space compared with shell boilers; and (3) unit construction enables its erection in an existing building. Firing can be by mechanical stoker, or automatic oil or gas burner. Outputs of 500 kW and upwards are available (corresponding to 900 kg of steam and upwards per hour from a steam boiler).

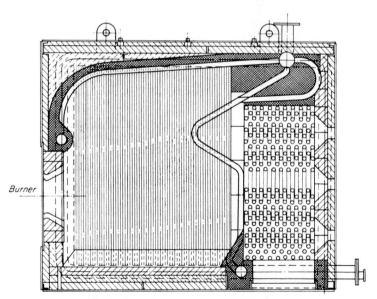

FIG. 30. Corner-tube boiler

Another design is the Corner-tube hot-water boiler illustrated in Fig. 30. Its water tubes are arranged to make a completely welded structure strong enough to carry the heating surfaces, the settings, and the insulating casings. The tubes form a large water-cooled combustion chamber. Fuels, firing and output are the same as in the previous type, and the Corner-tube is similarly economical in steel and floor space.

Boiler Instruments

It is the business of the boiler maintenance personnel to provide as much steam as is required, using the minimum fuel. The efficient use of fuel can only be achieved after considerable experience and training; but it is made much easier when the boiler is adequately equipped with instruments. In the next three paragraphs the operation of a natural-draught boiler with the bare minimum of instruments is described. In later paragraphs a brief account is given of additional instruments and their value in helping to secure efficiency.

As precautions against a burst boiler, a *steam-pressure gauge* and a *safety valve* are required by law; also a *gauge glass* showing the water level, and a *low-water alarm*. If the pressure gauge shows that the pressure in the saturated-steam space is falling, the staff know that steam is being used more quickly than it is being generated, and they increase the rate of burning by opening the dampers and so admitting more air to the fuel. If the pressure of steam rises, they reduce the rate of burning by closing the dampers. If the pressure exceeds the limit of safety for the boiler, steam escapes through the safety valve.

Water must be admitted to the boiler at such a rate as to keep the water in it at the working level. All water, with the exception of that from condensed steam, contains solids in solution, but no solids escape with the steam, so periodically the "blow-down" valve at the bottom of the boiler is opened through which water with entrained solids escapes under the pressure of the steam above. Other operations undertaken at longer intervals are the stoppage of leaks in the brickwork setting of a Lancashire boiler; removal of scale from inside the boiler when it is shut down; and removal of ash from fire tubes, economizers, super-heaters, the base of the chimney and other places of accumulation in the flues.

It is quite possible to operate a boiler with no instruments apart from a steam pressure gauge and a water gauge but, unless further measurements of one sort or another are made, it is difficult for the outsider either to obtain or even discuss fuel efficiency.

The *rate of inflow of water* and the *output of steam* are obviously important. If they are expressed in the same units, kg per second, and no steam is being wasted about the boiler, these measurements will be equal.

The *output of steam* can conveniently be measured by the method of Fig. 31. An alternative way is to determine the drop in pressure when the steam passes through an orifice of, say, half the diameter of the steam pipe. A flanged joint is made in a straight part of the pipe well away from bends or other restrictions of flow, and an orifice plate is inserted. Small pressure-lead pipes are inserted into the steam pipe at positions, which should be determined by formulae, on either side of the orifice. The pressure leads are connected to a manometer, which may be a dial-reading or recording instrument. The velocity of the steam through the orifice is approximately proportional to the square root of the pressure drop. Accurate formulae are given for calculating the flow of steam in kg per second, making corrections for the temperature and moisture content of the steam. If the steam is not superheated, and contains liquid water, the moisture content may be determined from the fall in temperature when a small quantity of the steam is allowed to expand to a lower pressure.

Manometers are also used for measuring the *draught*, i.e. the pressure differences drawing air through the fuel bed and through the boiler. The total draught is the difference in pressure between the air under the grate and the gases about to pass into the chimney. As this is usually small its measurement requires sensitive gauges. However, it is a useful index of the rate at which fuel is being burnt.

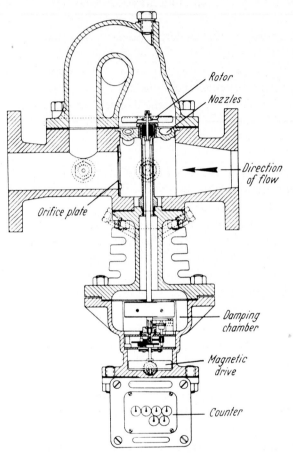

Fig. 31. Water or steam meter

There are a number of places in boiler systems where useful information may be gained by measuring *temperature*. Thermometry is not an easy subject, and numerous precautions are necessary if the measurements are to be reliable. The following types of thermometer are used: for the temperatures of steam, water, and air, either mercury-in-glass or dial-reading mercury-in-steel thermometers, or thermocouples which operate dial-reading or recording millivoltmeters; for the flue gases, thermocouples or mercury-in-steel thermometers; for the furnace itself, various forms of pyrometer for measuring the light or heat radiated from a convenient point in the furnace.

Industrial Boilers

The *consumption of solid fuel* can only be found by the laborious method of weighing or, less accurately, by estimation from the volume used. The test of whether fuel is being used most efficiently is the "equivalent evaporation" in kilogrammes of steam generated per kilogramme of coal burnt (see Table 16). If the calorific value of the fuel is also known, it is quite a simple calculation to arrive at the *thermal efficiency* which can be expressed as

$$\frac{\text{heat in steam}}{\text{heat content of fuel burnt}}$$

(see right-hand column of Tables 16 and 17).

A *heat balance* (i.e. a statement of where the heat goes) can be drawn up when the following are known: (a) the thermal efficiency, (b) the heat in the flue gases escaping up the chimney, (c) the potential heat in the unburnt combustible gases and (d) the potential heat from the carbon content of the clinker and ashes. For the measurement of (b), (c) and (d) different forms of *calorimeter* may be used, but as a rule chemical analyses are made and formulae applied. Table 17 is an example of a heat balance, from a Lancashire boiler, under very good conditions of operation.

TABLE 17. *Example of a heat balance*

A Lancashire boiler, with integral superheater, burning bituminous coal of 27·2 MJ kg^{-1} (as fired), gave 6.58 kg of steam at 284°C and 9·8×10^5 Nm^{-2} for each kg of coal. The water-feed temperature was 9·4°C.

	% of heat in fuel
A: Total heat in the 6·58 kg of steam = 19.6 MJ	
B: Calorific value per kg of coal = 27.2 MJ	
Therefore Thermal Efficiency (100 A/B)	72†
Heat lost in flue gases: Water vapour† 5 %	
Other gases 15 %‡	
Total	20
Potential heat in unburnt gases	2‡
Potential heat of the carbon in ash and clinker	3
Radiation losses etc., to make up 100%	3
	100

Notes: † Water vapour produced from the moisture in the fuel and from combustion of the hydrogen in the fuel.

‡ These three items are the ones most likely to vary under different firing conditions.. The figure for radiation losses depends very much on the design of the boiler, but should be constant within 1 for 2 per cent for any given boiler.

Carbon dioxide

The composition of the flue gases is in itself an index of the efficiency of combustion. Combustion is inefficient either (a) if too little air enters the furnace, when the flue gases contain too much combustible matter (usually hydrogen, methane, carbon monoxide and smoke) or (b) if an unnecessary amount of excess air is entering, when the flue gases contain too much oxygen and nitrogen. The mass of air actually needed for combustion is about 12 times the mass of coal burnt, and the mass of the hot flue gases is then 13 times. If, for example, 100 per cent excess air is admitted to the furnace the mass of the flue gases will be 25 times the mass of coal burnt. As a result the temperature rise of the flue gases will be only about half what it might have been: the rate of heat transfer from the gases to the water and steam will be less than half, and the wasted heat carried out of the chimney top will be about twice as much as need be. Thus the inefficiency of type (b) above can be very serious; but some inefficiency is unavoidable because with no excess air it is impossible to burn nearly all the fuel. The best results are obtained with 20–50 per cent excess air (see Fig. 33).

A good way to check the proportion of excess air is to measure the percentage of carbon dioxide in the flue gases. Air contains approximately 21 per cent by volume of oxygen. If combustion could be completed with no excess air, and if coal may be regarded as pure carbon, the flue gases would contain 21 per cent of carbon dioxide. With about 40 per cent excess air, and coal of typical composition, the flue gases contain 12–14 per cent of carbon dioxide, and this corresponds to the best attainable efficiency.

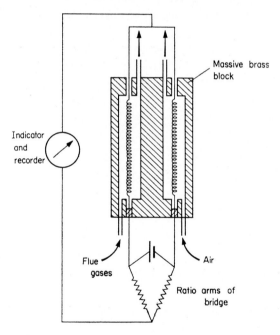

Fig. 32. Principle of the heat-conductivity carbon dioxide recorder

The percentage of carbon dioxide in flue gases may be measured by passing a sample of the gases through an absorbent which removes carbon dioxide, and automatic recording forms of this chemical apparatus are in use. Simpler equipment, such as that illustrated in Fig. 32, records the heat conductivity of the flue gases which decreases as the percentage of carbon dioxide rises. The scales of all such instruments are calibrated in terms of carbon dioxide because this is undoubtedly the key constituent of the flue gases.

Smoke as an index of efficiency

Where coal is burnt, the efficiency of combustion may also be judged from the smoke. The upper curve of Fig. 33 represents the total heat loss in a particular furnace; i.e. the heat wasted in warming any excess air, added to the heat which could be produced by burning all the combustible matter in the flue gas. This curve never reaches the zero line, because in a limited time it is impossible to make every gram of fuel combine with oxygen unless a very large excess of air is supplied. The curve reaches a minimum when the air supplied is between 20 and 40 per cent above the theoretical requirement of the coal. As can be seen from the horizontal scales of Fig. 33, the flue gases will then contain an appreciable

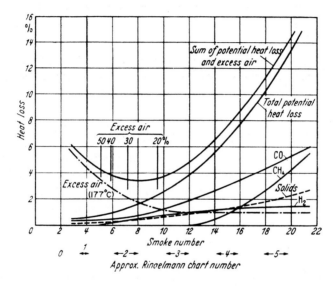

Fig. 33. Heat losses—full load

amount of smoke. The two horizontal scales are two different ways of measuring smoke, but the appearance of the smoke at the chimney top, when 20 per cent of excess air is used, is usually described as a light haze. If there is any more smoke than this, insufficient excess air is being used and efficiency is being lost. If there is a little less smoke, the efficiency is still very near the maximum. If there is no smoke, either all the coal in the furnace has turned to coke, or some special anti-smoke device is being used, or too much air is being admitted into the furnace. In the last case, the coal is being used wastefully, just as if smoke were being produced.

The boiler or furnace attendant needs to be shown continuously how much smoke the flue gases contain, prior to smoke-control devices. Figure 34 illustrates a photoelectric smoke indicator, for which a light projector and a photoelectric cell receiver are installed in the chimney at opposite sides. A remote indicator recorder and alarm bell are usually added to the equipment.

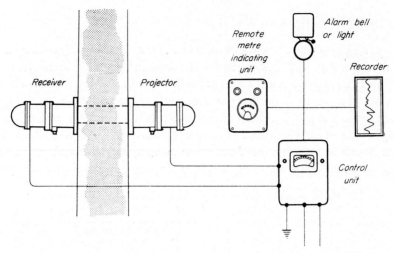

Fig. 34. Smoke-density-indicating equipment in conformity with BS. 2811–1969. (Courtesy of General Electric Co., Ltd. and marketed by Prat-Daniels (Stroud) Ltd.)

Alternatives to Coal

Small hand-fired boilers when burning coal will produce smoke, especially after the fire is raked or refuelled, and whenever the boilers are worked beyond their rated output. In theory, but apparently not often in practice, smokeless combustion can be achieved if coal is added in small amounts at a time, and provision is made for extra secondary air to be admitted, for example through nozzles above the fuel bed or through openings in the fire door. Generally, however, it is better to burn coke, low volatile coal, or anthracite, large charges of which can be given at times when the demand for steam is not too great. More and more use, too, is being made of oil and gas as fuels for the small boiler.

Alternatives to coal and other hard fuels may be sought on grounds only indirectly connected with outdoor pollution. For example, certain photographic processes, pharmaceutical laboratories and scientific "standards" rooms require dust-free atmospheres. In food handling and preparation, also, stringent standards must be kept, and gross pollution from refuelling or cleaning of solid fuel ovens in bakeries and ranges for fish and chip frying would be intolerable. In all these cases only the refined fuels such as electricity, gas and oil are acceptable.

When coal is the fuel, hand firing of industrial and commercial boilers rated at over 16 kW is virtually a contravention in Britain of the Clean Air Act 1956 (see Chapter 16). The lack of a mechanical stoker is evidence of failure to take "all practicable steps to prevent or minimize the emission of dark smoke". The firm must choose between the various smokeless fuels or between the various mechanical stokers several of which are described in the next section. Here are a few general principles for its guidance.

The nature of the load, the fuel, the boiler and the mechanical stoker or firing appliance must all be considered together. A fluctuating load requires a system that can respond quickly when a higher burning rate is required, and boilers and appliances of many sorts exist which have this property. A plant which uses forced draught and a short chimney invites trouble from grit emission, because free-burning small coal is readily carried into the flue gases and discharged from the chimney top; here a sprinkler stoker, for example, would be very unsuitable. Nuisance from grit is in many areas a bigger industrial problem than smoke. On the other hand, oil firing, which eliminates grit, will produce another nuisance, acid smuts, if the appliance is not well designed. Acid smuts, up to 7 mm in diameter, are discharged if (1) the flue gases become excessively cooled as in metal ducts and stacks, with the result that water condenses and becomes acidified by sulphur oxides, and soot is deposited, and (2) the deposits become large enough for fragments to be broken off and carried away by the flue gases. The acid smuts damage paintwork, stain light-coloured absorbent surfaces, make holes in nylon clothing, and attract public attention perhaps out of proportion to the real damage they cause. Finally, the selection of mechanical stokers for a particular boiler installation is only partly a matter of personal choice because it is usually restricted by the size, design and layout of the boiler house, the most economical fuel available, the capital cost, and opportunities for servicing and maintenance.

Mechanical Stokers

The Chain Grate Stoker (see Fig. 35) consists of cast-iron links made into an endless belt the width and length of which are determined by the size of the boiler. As the moving grate travels into the furnace an even layer of coal is taken on to its surface from a hopper; the thickness of the layer is controlled by a "guillotine" door. The output of the boiler is controlled by a combination of the speed of the grate and the thickness of the fuel bed both of which can be varied. The primary air which passes upward evenly between the links of the part of the chain grate within the furnace, is usually provided by means of forced draught. Secondary air can be introduced over the fire bed and regulated to give complete combustion of the volatiles. The ash that falls from the back of the grate

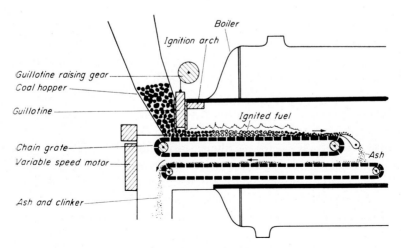

FIG. 35. Chain grate stoker

can be handled by an ash conveyor fitted below the grate. This brings the ash to the front of the boiler where it is automatically removed.

The Chain Grate Stoker is suitable for operation with nearly all types of shell boilers including Economic, Lancashire, and the various forms of multi-tubular boilers as well as brick set furnaces. It generally uses washed small coal, and is capable of burning efficiently the fuels of lower calorific value and high mineral content. A deposit of ash is an advantage on the grate links since it protects them from excessive heat.

Coal is fed to the *Coking Stoker* (see Fig. 36) by a reciprocating ram fitted beneath the coal hopper, and it moves forward along the grate by a combination of the reciprocating movements of the grate and the feeding of fresh coal from the front by the ram.

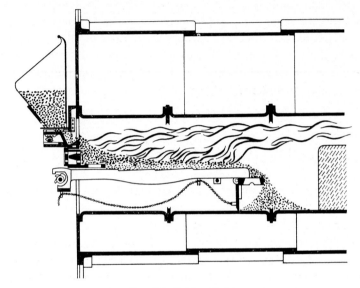

FIG. 36. Coking Stoker

At the same time the ash is moved to the rear of the grate. The Coking Stoker gets its name from the fact that the volatiles are largely released from the coal whilst it is being "coked" prior to its final combustion. Because the volatiles will pass through a hot incandescent coke bed they are consumed at the front of the grate, and this eliminates the emission of smoke. Combustion air flows through the grate and high burning rates can be obtained on short lengths of grate. Coking Stokers frequently operate under natural draught only but where this is not sufficient for the maximum load an induced draught fan may be fitted.

This stoker is widely chosen for shell boiler installations because of its adaptability to a wide range of coals and because of the high efficiency and flexibility that can be achieved along with smokeless operation under automatic combustion control. Like the shell boilers it is rather slow to respond to changes in demand for steam.

The Underfeed Stoker (see Fig. 37) is particularly suitable for smaller boilers such as the vertical water tube and for sectional and central heating boilers. Nevertheless it can also be used for shell boilers of either the single or twin flue variety. It functions most successfully on free-burning, non-caking graded coals of relatively low fines content and combines efficiency with smoke-free operation.

As the name implies the raw coal is forced upwards from the bottom of a retort in which the fuel is burned. The coal is conveyed from the hopper by means of a feed worm as illustrated.

As fresh coal is forced up the retort the coal at the top in the last stages of combustion falls over the sides of the retort and combustion is completed before the ash and clinker are finally removed through the furnace door.

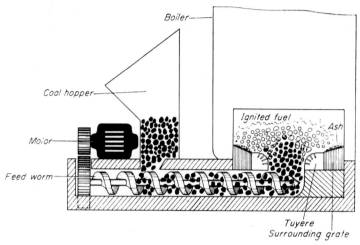

FIG. 37. Underfeed Stoker

The Underfeed Stoker operates under forced draught conditions and automatic control equipment is available so that a correct fuel/air relationship is maintained during load fluctuations. Highly caking coals are not suitable as they restrict the air flow. Air for combustion is supplied through the tuyères and the volatiles given off by the coal must pass through the ignited fuel thus eliminating smoke. A number of boilers can be linked together if required under a system of automatic control.

The Sprinkler Stoker (see Fig. 38) was the first of the mechanical stokers for shell boilers. In fact, an early type of stoker consisted of a spring-loaded shovel to imitate hand firing. In the diagram the coal is fed from a hopper by means of a feed control to a distributor, consisting of a set of revolving paddle blades, so that the coal is distributed continuously but in small quantities over the surface of the grate, thus maintaining a thin and even fire bed. A special feature is the rapid response of the firing rate to boiler demand. Both primary and secondary air may be used and the grate is normally of a fixed type although self-cleaning travelling grates may be fitted. Cooling of the firebars with steam or air is generally necessary.

Automatic control can be simply applied. The flexibility of the Sprinkler Stoker and its ability to burn coals of a highly caking or clinkering nature are its principal attractions.

Grit emission is kept to a minimum by careful selection of the coal to be used and by attention to operation of the stoker.

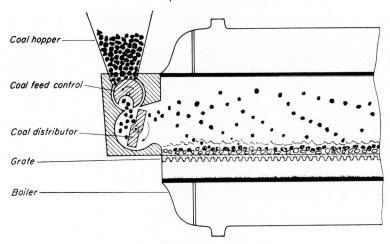

Fig. 38. Sprinkler Stoker

Pulverized Fuel

Powdered coal and charcoal were first tried in 1806 by French engineers in an attempt to produce an effective internal combustion engine. In 1837 John Samuel Dawes, a West Bromwich Ironmaster, was granted a patent for a method of firing blast furnaces by injecting pulverized fuel with the air through the tuyères. The use of pulverized coal is almost entirely restricted to large boiler plants and power stations burning 20 and more tonnes per hour.

The coal is ground to such fineness that 80 per cent will pass through a 200 B.S. sieve, whose holes are approximately 0·075 mm (75 μm) square. When thus pulverized the coal can be carried in an air stream through pipes and burned like a gas without the use of a grate. To supply small plants coal may be pulverized at a central depot by a merchant and delivered in special tank wagons, but more usually the grinding is carried out near the plant. Hot air passing through the grinding mill vaporizes excess moisture, carries the fuel to the burners, and then serves as the primary air for combustion, see Fig. 39.

The removal of moisture is so essential to the pulverizing process that only coal of low inherent moisture are really suitable. Another important property is low mineral

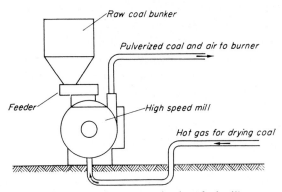

Fig. 39. Unit system pulverized fuel mill

Industrial Boilers

content. Inevitably the ash left after combustion is carried away in the flues and must be trapped before the flue gases are emitted to the atmosphere. This is obligatory in the U.K. Dust and grit removal is so costly (see Chapter 15) that only fairly high-grade coals low in mineral matter are suitable for pulverization. However, the development of mechanical coal cutting has much increased the amount of small coal available for grinding.

It is important to select the type of pulverizing plant which will give the lowest capital cost, be easily maintained and operated, and produce fuel having the required degree of fineness. The choice must vary widely with the types of coal available and their moisture content. Figure 40 illustrates a design of pulverized fuel burner.

Amounts of pulverized fuel consumed in Great Britain are given in Table 18.

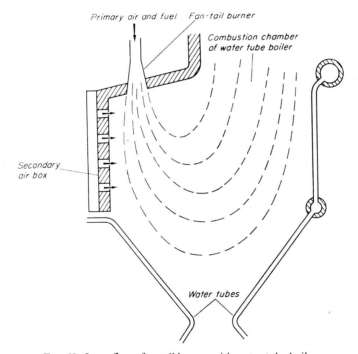

FIG. 40. Long flame fan-tail burner with water-tube boiler

TABLE 18. *Consumption of pulverized fuel in Great Britain*

	Millions of tonnes			
	1930	1938	1950	1960
Electricity undertakings (power stations)	0·94	2·22	7·80	34·97
Cement and other kilns	1·37	2·14	2·80	2·83
Collieries	0·30	0·65	0·80	0·43
Metal Manufacture	0·04	0·24	0·67	0·18
Other Industries	0·45	0·87	1·24	3·10

Source: Ministry of Power, *Statistical Digest*, 1960.

Boiler Availability

Water-tube boilers are designed to use the greatest possible amount of heat from the fuel. In a modern power-station boiler the flue gases usually pass between banks of boiler tubes, superheaters, economizers, and air preheaters. If the gases contain condensable vapours or solid particles in suspension, these may accumulate on the tubes and (1) seriously impair the transfer of heat from the gases to the water, steam or air within the tubes or (2) so choke the passages between the tubes that they cease to function properly. Power-station boilers should be able to remain on load for 6–12 months without being shut down for cleaning, but there have been cases of boilers shutting down within a few weeks because of (2) above. The effect has been to increase the number of power stations or, at any rate, the number of boilers necessary to satisfy the demand for electricity.

In Britain the investigations of a war-time Boiler Availability Committee revealed how boiler deposits are formed and how they may be kept within bounds. On tubes which may be reached by the flame, deposits known as "bird-nesting" occur as a result of the accumulations of coal ash, solidifying from the molten state. On tubes and other surfaces more removed from the flame, "bonded deposits" occur, which are particles of ash and grit immersed in a bonding medium of the sulphates or acid sulphates (mainly pyrosulphates) of sodium and potassium. These salts are formed from the sulphur, sodium, and potassium present in coal; they have relatively low melting points, and form sticky deposits on the cooler surfaces, to which particles of ash and grit will readily adhere.

"Phosphatic" deposits are similar in appearance to bonded deposits, but they are caused by the presence of phosphorus compounds distilled from certain coals; fly-ash particles are chemically attacked by phosphorus or its compounds and hard masses of phosphates are formed. Since the susceptibility of a boiler to bonded or phosphatic deposits depends very much on its design, and since there are coals containing very different proportions of sodium, potassium, and phosphorus, the deposits can be reduced by the allocation of coal suitable to each boiler.

Deposits on economizers are due to fly ash in the presence of sulphur and phosphorus compounds. Air heaters suffer serious deposits if the heating surfaces are so cool that liquid from the flue gases collects on them, and it should be noted that the presence of small quantities of sulphuric acid may raise the dew point to 150°C or more. Evidently coals containing a high proportion of sulphur are to be avoided if possible where air-heater trouble is prevalent.

Pulverized-fuel installations are most likely to suffer from re-fusion deposits (bird-nesting) on the boiler tubes and super-heaters, because of the quantity of fly ash in the gaseous products of combustion; the trouble can be reduced by designing the combustion chamber to have the maximum amount of cooling by water tubes embedded in the walls. Apart from the accumulation of dust in the economizers and air heaters, pulverized fuel installations are almost immune from all other forms of deposit. Boilers with mechanical stokers and retort stokers do not suffer seriously from refusion deposits unless very fine coal is burnt, but they are susceptible to the other forms of deposit unless suitable precautions are taken. The high temperatures in the fuel beds of some retort-stoker boilers make them particularly liable to trouble if the coal contains much phosphorus.

Oil-fired boilers burning heavy fuel oil suffer from deposits in the super-heaters and the air preheaters, and, more serious, corrosion of low temperature metal surfaces in the air heaters. Both troubles are consequences of sulphur trioxide formed during combustion of sulphur in the fuel. The corrosion can be reduced by the addition of ammonia to the flue gases or of powdered dolomite to the burning fuel. The quantities needed are quite large, about 1 per cent of the weight of fuel.

Soot blowing

The efficiency of both large and small water-tube boilers is improved if the heating surfaces can be kept free from deposits while the boiler is on load. This ideal can be to some degree realized by the intermittent spraying of the surfaces with jets of steam or water. Large boilers usually have steam jets built into positions where they are likely to be most useful, and steam is blown through them at regular intervals of about 8 hr. The operation of soot blowing may cause the emission of particles to the atmosphere together with some smoke, since the deposits usually contain a proportion of combustible matter. When there are suitable grit arrestors in the system, however, the emission may be quite negligible. Allowances are made in the Clean Air Act for soot blowing and flue cleaning (see Chapter 16).

Fluidized Beds

The use of fluidized beds in combustion processes is a recent development, although the technique has been used in the chemical and metallurgical industries for catalysis and heat treatment since the 1920s. Injection of a strong current of air into a bed of sand (from below) lifts the particles so that the sand bed behaves like a boiling fluid. It is found that in such a "fluidized bed" almost anything (including wet sewage sludge) can be successfully burned by maintaining the bed at a temperature of 700–900°C. The advantages are many: a lower temperature restricts formation of nitrogen oxides, the fluid nature maintains even heat distribution, heat transfer to the boiler pipes (immersed in the sand and thus reducing the overall size of the boilers) is more efficient (by a factor of about 4), and, by the addition of pulverized limestone to the bed, the formation of sulphur dioxide can be prevented. More than 90 per cent of the sulphur in fuel can be removed in this way thus allowing high sulphur content fuels (e.g. Appalachian coal in the U.S.) to be used.

CHAPTER 8

Industrial Furnaces

IN THE last chapter, consideration was given to the design and operation of boilers. A second major consumer of fuel for combustive processes is the furnace. Some data on different furnaces and fuels are collected in Table 19. The different types of furnaces are numerous; but can usefully be grouped under three main headings:

(1) *Substance separated from fuel and gases.* Examples: boilers, coke ovens, gas retorts, crucible furnaces, muffle furnaces, electric, gas and oil-fired furnaces, oil refineries.

(2) *Substance and fuel in contact.* Examples: blast furnaces, cupola or shaft furnaces, charcoal heaps and beehive coke ovens.

(3) *Substance in contact with flames or gaseous products of combustion.* Examples: hearth furnaces, reheating furnaces, pottery kilns, brick kilns, lime and cement kilns.

TABLE 19. *Fuel burnt in different types of furnace*

Substance heated	Furnace	Fuel
Water	boiler	coal
		coke
		oil
Coal	retort	producer gas
	coke oven	coke-oven gas, etc.
Petroleum	refinery	oil
Heavy clays	continuous	coal, producer gas, oil
Glass	regenerative	gas, oil
Pottery	intermittent	coal
	continuous kiln	producer gas
Limestone	lime kiln	coal
Cement slurry	rotary kiln	pulverized fuel
Iron ore	blast furnace	coke
Iron and steel	regen. hearth,	gas
(melting)	shaft, crucible,	coke, gas, oil
	arc, induction	electricity
(heat treatment)	intermittent,	all fuels
	continuous	gas, p.f., oil
Non-ferrous ores and	smelter,	coal
metals	crucible	coke
Aluminium and	electrolytic,	electricity
magnesium	induction	

Group (1) Furnaces

Some typical examples of furnaces which are likely to be of general interest will now be considered. In group (1), boilers apart, coke ovens and retorts are the most important consumers of fuel.

Figure 41 shows how the fuel was burnt in the carbonizing plant of a gas works. Coke breeze from the hopper was fed into the gas producer. Producer gas passed immediately

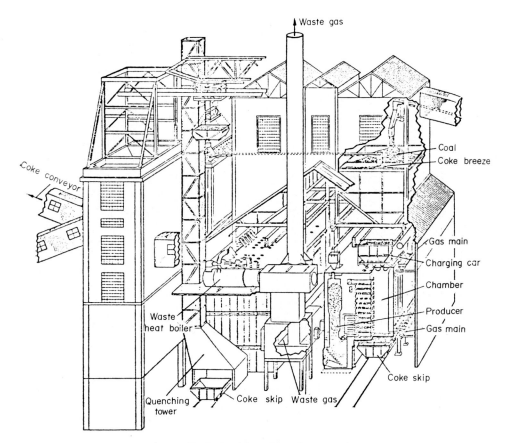

FIG. 41. Carbonizing plant of a gasworks

into the combustion space round the coal chambers. After combustion the waste gases passed through the waste-heat boiler into the chimney stack. (The waste-heat boiler was a gas-tube boiler similar to the economic boiler illustrated in Chapter 7, though without the main furnace tube. The steam was used to provide mechanical power or to generate electricity for use within the gas works.)

Although carbonization is no longer carried aut to produce town gas, Fig. 41 may be taken as representative of many carbonizing plants, though there are important variations in design and operation, some of which are discussed below.

Horizontal retorts

Although developed for and largely implemented in the carbonization of coal in gas works (see Table 20 below) and thus obsolete in England and Wales, the horizontal retort will be discussed here because of its similarity on a small scale to coke ovens. It is a tube of silica brick or fire-clay about 6 m long, 60 cm wide, and 40 cm high. From six to ten were arranged in one "setting" (Fig. 42). These were heated by producer gas from a

Fig. 42. Horizontal retort beds at gasworks

single, coke-using producer, admitted through about twelve openings to the combustion chamber which enveloped all the retorts in the setting. The air for combustion was preheated in *recuperators* where heat was transferred to it from the waste flue gases through 7·5 cm dividing walls of fireclay. Less than half the heat of the flue gases was recovered in this way, but up to three-quarters of the remainder could be used for raising steam in waste-heat boilers.

The retorts were charged about five-sixths full with coal of sizes up to about 2·5 cm. These were heated to about 1000°C for about 12 hr. The products of carbonization passed from the ends of the retorts usually into ascension pipes from which they travelled down dip pipes into hydraulic mains placed along the top of the retort setting, and here the processes of tar and liquor separation and gas purification commenced. The coke was discharged by pushers into cars (Fig. 43) where it was cooled by water.

Fig. 43. "Coke bus" for disposal of coke at gasworks

Coke ovens

The width of a horizontal retort is limited by the need for all the coal within it to be heated equally. There is no restriction on its height, provided the sides can be adequately heated. Coke ovens are very high retorts, a common size being 0·5 m wide, 4 m high, and 12 m long. Coke ovens in Britain are mostly employed in the manufacture of metallurgical coke, i.e. under conditions where sudden fluctuations in demand do not occur.

Coke ovens are built in batteries of 25–60, and each charge of 15–25 tonnes of coal per oven is heated for 12–22 hr, its final temperature being about 1000°C. This temperature can be reached using coke-oven gas, producer gas, or even blast-furnace gas, by the employment of *regenerators* instead of recuperators. The regenerators are chambers containing fire-bricks in the form of a three-dimensional chequerwork, and four regenerators are used with each setting of ovens. The hot flue gases are used to warm two of the regenerators, while the incoming gas and air are brought to the furnace through the other two. At intervals of about $\frac{1}{2}$ hr four flap valves are reset, and the functions of the two pairs of regenerators are transposed. (If coke-oven gas is the fuel it is not preheated as it is liable to decompose, forming carbon; only the air is preheated.) Although regenerators are a highly efficient way of preheating the gas and air, the flue gases still retain about half their heat, and can be used in waste-heat boilers before escaping to the atmosphere.

The coke from most coke ovens is still quenched by water, requiring about 3 tonnes of water per tonne of coke. In one British plant dry-cooling by inert gases is used. It is claimed that dry-cooling, though complicating the plant, improves the quality of the coke, and saves heat equivalent to about 3 per cent of the original coal.

Vertical retorts

Horizontal retorts and coke ovens are intermittent in operation, because production from each retort or oven must stop during charging and discharging. There are also intermittent vertical retorts or chambers but the vertical retort which operates continuously was much more common in Britain, as can be seen from Table 20.

TABLE 20. *Gas produced by carbonization in Great Britain*

	% of total gas made (1959)
Continuous vertical retorts	65
Horizontal retorts	18
Intermittent vertical chambers	12
Coke ovens	5
	100

The table divides the total gas made in 1959 by British carbonizing plant among the four main types of plant. Between the 1940s and the 1960s there was an increase in the production from continuous vertical retorts, and to a lesser extent from coke ovens and static vertical retorts (see below).

There are several designs of verticalretort settings, but typical dimensions for an individual retort are 8 m high, 200 cm by 25 cm in section at the top, increasing to 200 cm by 46 cm at the bottom to allow for expansion of the coal during carbonization and to reduce its tendency to stick to the sides and form arches across the retort. Figure 41 shows a section through an installation of the sixties in London, which had two parallel benches each of 24 retorts, and carbonized 336 tonnes of run-of-mine coal per day. The retorts were heated by producer gas from two separate batteries of four gas producers. This was burnt at the sides of the retorts, and air for combustion was admitted at suitable levels to distribute the heat. The incoming air was partly heated by passing it round the outside of the combustion chamber, and the flue gases were used in waste-heat boilers.

The coal for carbonization was admitted through a gas-tight hopper at the top, and the coal gas withdrawn from near the top. The coke was discharged at the bottom by a mechanical extractor, after being cooled by steam, which was blown through jets into the extractor box. The steam combined with some of the coke to make water gas (see Chapter 5) and this passed up the retort and mixed with the coal distillates. The effect of the steam was that heat which would otherwise have been dissipated on cooling the coke was converted into gas. In the installation, of Fig. A a dust-extraction plant removed dust arising from the discharge of the coke.

Industrial Furnaces

The advantages of continuous vertical over horizontal retorts are (1) continuous operation, which implies uniformity of treatment of all coal passing through, and uniformity in output and quality of gas, (2) smaller ground space of plant, (3) dryness and ignitability of the coke, and (4) avoidance of smoke during charging and discharging operations. The possible disadvantages are (1) elaborateness of the mechanism of the coke extractor, (2) limitation of size, and (3) tendency to stick when coal is used having different caking properties from those for which the retort was designed.

Static vertical retorts

In design these consist of a vertical externally heated silica retort of rectangular section, and each static charge of coal needs some 10 hr to be carbonized. Coke cooling was formerly in an integral chamber beneath the retort with the generation of water gas, but is now carried out in an ordinary cooling tower.

Electric furnaces

Some of their many applications were mentioned in the last chapter. In the steel industry the production of ferro-alloys is a major one and a suitable furnace is illustrated in Fig. 44.

In steelmaking, open hearth furnaces have been largely superseded by electric arc furnaces using high-purity oxygen injection and capable of output rates greater than 60 tonnes per hour.

Oil refineries

The world's petroleum has to pass through refinery processes of distillation, conversion, etc. (see Chapter 4). There are major amounts of gases and liquids to be dealt with and it is only by the closest attention to every item of equipment that odours from a variety of petro-chemical processes are kept under control. Oil refineries are also a fairly large source of emissions of sulphur dioxide for it is not viable to desulphurize all escaping gases. Hence the need for high chimneys which are a characteristic of modern refineries.

Another feature is the refinery flare which serves as a master safety valve in cases of equipment failure or operational emergency. Fuel gas cannot be released to air partly because of odour considerations, partly because of explosion hazards. Consequently the flare stack is provided with a pilot flame so that any released gas may be ignited. The simple burning of refinery fuel gas at the top of a stack, however, would involve the production of much smoke unless steps were taken to prevent it. This is done by the injection of steam close to the point of ignition or by the use of water sprays. A major difficulty has been to devise a control system to adjust the supply of steam and water to the amount of gas.

7 A.P.

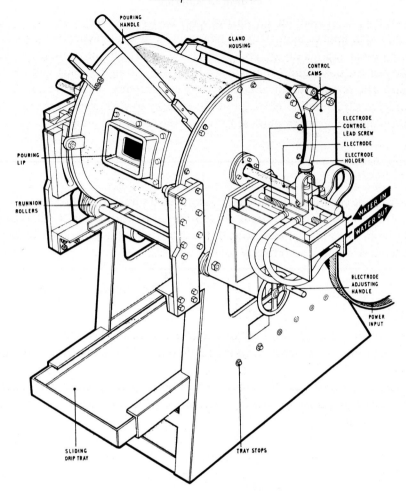

Fig. 44. Indirect arc furnace
(Courtesy of Cline Foundry Supplies Ltd.)

Group (2) Furnaces

Blast furnaces are the most important furnaces in group (2), where the substance and fuel are in contact. It will be realized, however, that the function of the fuel is not only to heat the ore, but to produce gases which react chemically with it. The principal use of blast furnaces in Britain is for smelting iron, but smaller water-jacketed blast furnaces are used for smelting lead.

The shaft of an iron-smelting blast furnace may be from 12 to 24 m high, and its diameter at the top 3 m, widening to 4 m at about 5 m from the bottom. It is illustrated in Fig. 45. It is filled to the top, and kept full during continuous operation, with the correct mixture of iron ore (iron oxide or carbonate), metallurgical coke, and limestone. The purpose of the limestone is to form quicklime on heating, which removes sulphur from the iron; it also acts as a flux, collecting the slag-forming material from the ore and the ash from the coke to form a liquid slag, and so separating them from the iron.

Air is forced through tuyères into the bottom of the shaft, where the temperature is over 1500°C, high enough to melt the iron and the flux so that they run down into the hearth from which they are tapped periodically. The air admitted is insufficient to burn the coke fully to carbon dioxide, and the gases which pass upwards through the central and upper part of the shaft contain a high proportion of carbon monoxide and no oxygen. The carbon monoxide withdraws oxygen from the iron ore, reducing it to metallic iron; a relatively small proportion of the carbon monoxide is oxidized to carbon dioxide. Thus the blast furnace performs two functions; it reduces iron ore to iron, and it separates the iron, by melting, from the mixture of minerals in which it then lies.

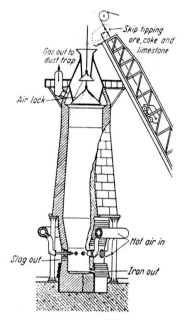

FIG. 45. Blast furnace

The gases which pass out of the top of a blast furnace still contain much carbon monoxide, nearly as much as there is in producer gas. In all modern installations this blast-furnace gas is dedusted and usefully employed as a fuel, in hot-blast stoves for preheating the air of the blast furnace, or in coke ovens, reverberatory furnaces, or gas engines. There are still some older blast furnaces whose gas is either led up stacks to the atmosphere or allowed to burn at the top of the shaft.

Irregular working or unsuitable coke may result in a "check" by the "arching" of the charge within the shaft. This is followed by a "slip" which causes a sudden rush of gas and the automatic opening of safety vents at the top of the furnace. When the gas pressure within the furnace is relieved in this way, the escaping gas is heavily burdened with dust and may be a serious source of complaint. Operating conditions should be such as to keep this "slipping" to a minimum. Double bell sealings or their equivalent should be provided for sealing the tops of furnaces and preventing the escape of gas during charging.

It is found that if minor slips are induced under controlled conditions, catastrophic slips (and the consequent large scale emission of noxious fumes) can be virtually eliminated.

Cupolas are a smaller type of shaft furnace (see Fig. 46), in which iron and steel are melted, usually for recovery from scrap metal, and cast into ingots or castings of special shape. The fuel is metallurgical coke, and the flux is usually limestone. The air blast may be used cold or preheated to 450–500°C. The hotter waste gases from the hot-blast

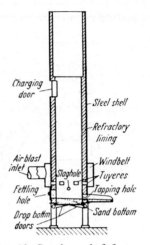

Fig. 46. Cupola or shaft furnace

cupola contain metallurgical fume from which they derive their striking colours. Both hot blast and cold blast cupolas are serious sources of grit and dust. Heavy smoke emissions, too, are inevitable if the scrap iron to be melted is contaminated with oil, grease, tar or other volatile matter; but smoke at the time of lighting can be avoided by the use of an industrial gas torch. The stocking area for metal, coke and limestone should have a hard paved surface, and the firing of adventitious ash and grit should be avoided.

Atmospheric Pollution from Furnaces in Groups (1) and (2)

All the furnaces so far discussed in this chapter should employ smokeless fuel. There are unavoidable emissions of smoke during charging and discharging where coal is carbonized in horizontal or intermittent retorts or coke ovens. The amount of smoke produced depends on the time the retorts or ovens are open for charging and levelling, usually 5–6 min in each coking period of 12–22 hr; when ovens were charged by hand the time was 15 min or more, and the emission of smoke was proportionately greater. Any other emissions of smoke at these plants are due to leakages or intermittent technical difficulties.

At the majority of coke ovens the hot coke is quenched with water, and this is a serious cause of pollution by grit and hydrogen sulphide. (Apart from sulphur dioxide which is discussed below, the chief smells associated with gas works used to be from tar and ammonia-liquor wells, from spent oxide being revivified, and from purifier boxes being emptied. If tar was distilled and ammonium sulphate was prepared at the works, there would also be the smells of hot pitch, phenols, and pyridine.)

From furnaces where coke is burnt, the principal forms of pollution are sulphur dioxide and particles of grit and ash. The sulphur dioxide generated may be taken as about 3 per cent by weight, on an average, of the coke burnt; not quite all of this reaches the atmosphere because a little is absorbed within the flue system. At gas works about 15 tonnes of coke were burnt for every 100 tonnes of coal carbonized; the emission of sulphur dioxide was about 2·7 per cent of the coke burnt or about 0·4 per cent of the coal carbonized. Since as much as a million tonnes of coal per year could be carbonized at a large gas or coke-oven works, and since the stacks were often a bare 30 m high and near to buildings of similar heights, the concentration of sulphur dioxide in the immediate neighbourhood was often considerable. Until recently, such proximity of pollution-producing industries to houses (Fig. 47) was not regarded as an important factor in

FIG. 47. Proximity of houses to industry (gasworks)

planning policy and resulted in high levels of toxic gases such as hydrogen sulphide being present for long periods seriously affecting the health of the local inhabitants (especially children).

Metallurgical coke is frequently made from coal which is low in sulphur content, and when it is burnt as little as 2·0 per cent of its weight may be emitted as sulphur dioxide.

It is hard to give a generalized estimate of the weight of grit and ash emitted from the various coke-burning furnaces which have been described. Only a small proportion of the particles can settle out in regenerators, recuperators, or other parts of the flue system, and particles may be emitted up to 3 or 4 per cent of the weight of coke burnt. A more usual proportion would be about 0·5 per cent. Grit from built-in producers causes blockage of flues and serious abrasion of the expensive brickwork of retorts and coke ovens.

Group (3) Furnaces

Steel industry

Open-hearth furnaces are an example of group (3), in which the substance is in contact with the gaseous products of combustion (see Fig. 48). They are used for melting steel, purifying it, and alloying other substances with it. The fuel burnt is usually oil, but

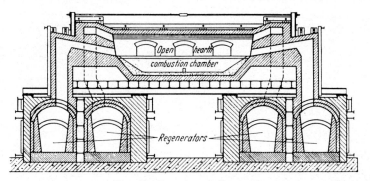

FIG. 48. Open-hearth furnace

sometimes producer gas from coal, a long, luminous flame being required above the bath of molten metal. Both producer gas and air for combustion are preheated in regenerators similar to those used with coke ovens. Atmospheric pollution is similar to that from coke ovens, with the addition of smoke if insufficient air is used in proportion to the producer gas.

Reheating furnaces are used for the heating of steel and other metals. The fuel used may be coal, pulverized coal, gas, oil, or electricity, and the furnace may be intermittent or continuous. In the intermittent furnace a batch of material is heated to the required temperature, when its place is taken by a second batch. In the continuous furnace the material is moved from the coolest to the hottest zone at a suitable rate; cold material is continuously being put in at one end and hot material taken from the other. All types of steel reheating furnaces are in group (3).

Clay industries

Numerous examples of furnaces in group (3) are provided by the clay industries, which make (a) refractory bricks of fireclay, silica, and other materials for industrial use, (b) building bricks, tiles, and pipes, and (c) china, earthenware, glazed tiles, porcelain, and many similar articles. The method is much the same in every case. Wet clay is moulded to the required shape, and heated, gradually at first to about 120°C to evaporate the free water, then more strongly to 900°C to drive water out of the molecules of clay, and finally to the baking temperature between 1200 and 1450°C.

Hand-made bricks were at one time fired exclusively in clamps, great skill was required both in setting and burning the bricks by this method. The simplest brick kiln, with a permanent structure, is the up-draught or Scotch kiln, of four upright walls forming a rectangular chamber usually about 8 m long by 5 m wide and 4 m high. The two end walls are removable to facilitate filling and emptying. Along each side are fire holes about 40 cm wide and 0·6–1·0 m high, lined with firebricks.

The down-draught kiln (see Fig. 49), whether circular or rectangular, is the most efficient and satisfactory of intermittent kilns for firing all kinds of clay products. In 1955

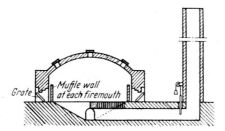

FIG. 49. Down-draught kiln

there were an estimated 5000 in operation, mainly hand fired. Hand firing has been almost eliminated under the Alkali Act and the total number of downdraught kilns had fallen to 424 in 1973. In a typical coal-fired kiln, fires would be lighted in the 10 or 18 grates, not much bigger than domestic grates though taller in shape, round the outside. Smoke and gases from the fires pass up over a wall towards the crown of the kiln and down through the ware into bottom flues, beneath the floor of the kiln, then through an underground main flue to the chimney stack, usually separated from the kiln and about 9 m high.

The entire kiln is cold to begin with, and when the fires are made of bituminous coal much of the volatile matter in the coal escapes as smoke from the chimney. Smoke emission during the "water-smoking" period (up to 120°C) can be eliminated by using low-temperature coke, or anthracite. If these fuels are not available in quantity, smoke can be reduced by using steam coal, or a mixture of coal and coke. Other precautions are (a) to shorten the time required by starting with the ware as dry as possible, (b) to charge the fuel in small quantities, alternate fires at a time, and to put the fresh coal at the front of the fires, (c) to admit secondary air, preheated to about 500°C, over the fuel bed, (d) to remove ash and clinker from one fire at a time, (e) to keep the kiln and flue structures in good repair, (f) to increase chimney heights and (g) to convert to oil firing. Mechanical stokers go a long way towards meeting these problems, and the underfeed stoker is now finding increasing use in down-draught kilns (see Fig. 37). Coal and oil firing are being replaced by natural gas and smokeless petroleum gases (butane and propane).

During the later stages of firing, when the kiln is very hot, it is generally necessary to use long-flame bituminous coal to ensure that no part of the ware is overheated. Correct use of the fire doors and of secondary air controls will ensure smokeless combustion. Insufficient air will cause heavy smoke, but too much air may spoil part of the ware by overheating.

For the very highest temperatures to be reached and maintained, excess air must be avoided and smoke is usually emitted, especially while refuelling is in progress. Another difficulty during the later stages is that some ware has to be heated in a reducing atmosphere, to convert ferric oxide to ferrous oxide and produce a blue colour rather than red. For a long time there appeared to be little hope of reducing the emission of smoke in these circumstances, but smoke can now be avoided by using tertiary air. This is drawn through special flues under the kiln where it is heated; it then mixes with the waste gases and the smoke is burnt within the flues before it can pass up the chimney. The heat of combustion of the smoke is not entirely wasted, since it helps to raise the temperature of the bottom of the kiln.

Intermittent kilns, such as those which have just been described, are still considered the best for producing silica goods, salt-glazed pipes, large fireclay blocks, some kinds of tiles, and materials of which the required output is small. For all other products *continuous kilns* are preferable, and they can provide all the necessary conditions for the elimination of smoke and most of the fly ash. They have been used in the brick industry since 1858. There are many types of continuous kiln, in all of which the flue gases are used for heating the cold goods, and the heat of the finished goods, during their cooling, is used for preheating the air. All stages of drying, heating and cooling go on concurrently in the kiln, different units of the ware going through each stage in succession. The thermal efficiency of continuous kilns is high; again smoke emissions can be obviated by careful control or conversion to smokeless fuels such as butane or natural gas.

Lime and cement kilns

Unlike pottery, lime and cement are products whose cost is greatly affected by the fuel economy of their manufacture. Much attention has therefore been paid to the fuel efficiency of the kilns in which they are made.

Relative cement-making capacities (million tonnes/yr) in different countries in 1958 were: Britain 13, U.S.A. 53, U.S.S.R. 25, Germany 20, Japan 13, Italy 11, France 11.

Lime, i.e. calcium oxide, is made by heating limestone, i.e. calcium carbonate, to temperatures of 1000–1100°C. The reaction is

$$CaCO_3 \rightarrow CaO + CO_2.$$

Lump limestone is "burnt" in continuous vertical kilns, heated formerly by coal or producer gas but nowadays mostly by petroleum oils. Each lump of limestone is in the kiln for about 6 hours, gradually falling towards the hot zone as the finished product is withdrawn from the bottom of the kiln. Lime is also manufactured from powdered limestone in rotary kilns similar to those which will now be described in connection with the manufacture of cement.

Portland cement is made by heating a mixture of lime and clay to about 1300°C, forming complex calcium silicates and aluminates which fuse and produce clinker on cooling. The clinker is ground to the familiar grey powder.

The rotary cement kiln illustrated in Fig. 50 is a steel tube, lined with alumina bricks, about 3 m in diameter and 30–120 m long. It is steadily rotated at about 1 rev/min. The kiln is slightly inclined so that materials fed in at the upper end travel slowly to the lower end in 2 to 3 hours. Pulverized coal and air are admitted at the lower end, and the products of combustion travel up the kiln and on into a chimney stack (via appropriate control devices).

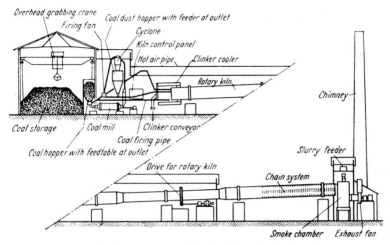

Fig. 50. Rotary cement kiln fired by pulverized coal

At the top of the kiln the mixture of lime and clay enters as a wet slurry. The inside of the kiln carries numerous slack chains which fall across as it rotates and bring the maximum quantity of slurry into contact with the flue gases. Great care must be taken to ensure that the slurry does not become dry too near the top of the kiln. If this happens, much of it crumbles to a powder and is carried away by the flue gases. This is both wasteful and a cause of serious nuisance in the form of deposited atmospheric pollution.

Figure 50 and the above descriptive paragraph may be regarded as typical of all but the newest cement works. Instead of using a slurry containing 40 per cent water, the present trend is to form the raw materials into nodules having no more than 20 per cent of water and to remove this water by heat exchange without permitting direct contact with the flue gases. The rate of dust emission must be substantially less with this type of kiln and its use is likely to extend. Many kilns are being converted, also, from coal to oil firing.

The flue gases are normally passed through cyclones, which recover much of the escaping powder and reduce the pollution, but they are quite inadequate to prevent the emission of grit when the kiln is operating near the maximum capacity. Electrostatic precipitators are better, but they need to be kept in really good order, by providing kiln and precipitator in duplicate to permit repairs to be carried out. Chimneys up to 120 m high are in use to protect the immediate neighbourhood from nuisance. More dust-control devices such as multi-cyclones and electrostatic precipitators are installed in many cement works. These are described in Chapter 15. Unfortunately the need for fuel economy makes it important to remove as much heat as possible from the flue gases before they escape, and the risk of drying the slurry too soon is always present.

As well as the high level dust from chalk, limestone, clay or pulverized fuel, there is another nuisance—cement at ground level. This arises from grinding, packing and loading cement in bulk on to trucks.

Atmospheric Pollution from Furnaces in Group (3)

Smoke in the steel industry

It is usually necessary to keep the hot metal in a reducing atmosphere to prevent damage by oxidation and, since this atmosphere is produced entirely by the gaseous products of combustion, it is necessary that the fuel should be incompletely burned. When the fuel is coal, it follows that a great deal of smoke is liable to be produced, especially if the stoking is by hand, the fire grate is unenclosed, and there is no provision of controlled secondary air. The production of smoke tends to be greatest when the furnace or the charge is cold, but much can be done to reduce smoke by skilful stoking designed to keep the flame conditions steady within the combustion chamber.

It has frequently been claimed that the presence of smoke in the combustion chamber is an advantage. Perhaps this is because smoke serves as a convenient indicator that no oxidation is taking place; the conditions of temperature and oxygen supply at which steel oxidizes are probably about the same as those at which smoke burns. But it is quite fallacious to state that the only reducing atmosphere is a smoky one. In the blast furnace the atmosphere is strongly reducing, yet a smokeless fuel is used. Intermittent and continuous reheating furnaces are already in use in which the fuel is oil or producer gas, and heat is conserved in regenerators. The place of smoke as an indicator of reducing conditions is taken by instruments which record the volumes of fuel and air entering the furnace.

Sulphur dioxide and grit

As always, sulphur dioxide is produced in proportion to the amount of sulphur in the fuel. Much of this may be emitted to the atmosphere, unless sufficient waste-gas treatment (e.g. scrubbing) is employed to remove it. An alternative way to reduce sulphur dioxide emissions is to pretreat the fuel to remove sulphurous impurities (e.g. by washing).

The amount of grit and ash emitted from group (3) furnaces depends on the fuel and conditions of draught, tending to be particularly large from some of the metallurgical reheating furnaces. Unless there are large particles, for example those emitted where pulverized coal is burnt without efficient grit arrestors, the complaint will usually be of "smoke". There is much to be said, however, for distinguishing carefully between the different forms of particles. In the past, in a district whose atmosphere was already badly polluted there was a noticeable tendency for new and perhaps unnecessary pollution to appear.

Summary

The above discussion outlines the pollution problems presented by the industrial use of the combustion process. Increasingly, air-pollution-control devices which eliminate both particulates and undesirable waste gases are being employed; partly in compliance with the Clean Air Acts and partly in the context of public awareness of pollution and the environment and a conscious effort on the part of the politician, industrialist and the general public alike to reverse the trends of the early part of the twentieth century culminating in the great smog disasters of the early 1950s. (Removal mechanisms are dealt with in Chapter 15.)

CHAPTER 9

Domestic Heat Services

FROM the Dark Ages until Tudor times, most of our ancestors lived in one roomed huts or cottages. They were heated in winter by a central fire of sticks or logs, under a hole in the roof; so were even the baronial castles and Norman manors. This arrangement made good use of the heat obtained from the fuel, but whenever fresh logs were added, the room must have been filled with smoke. There is no doubt that the first smoke problem which faced the world was that of removing smoke from inside a dwelling; charcoal was the only smokeless fuel available in those days, and it must have been in considerable demand.

The problem of indoor smoke was solved by the Romans with a form of central heating. They built villas on raised floors of concrete and tiles, and made the fire underneath the floor. In this way they could burn even coal without discomfort. Their technique was forgotten in the Dark Ages, and it was not until the nineteenth century that modern central heating methods were introduced.

There was another internally smokeless form of house heating which was used in pre-christian Europe. The fire was made in an earthenware brazier outside the dwelling, and brought indoors when the smoke subsided. There was also a type of closed stove for baking and cooking, but the heating stove, built within a room and having a flue pipe built into the wall, did not appear until the eighth century. The early closed stoves were made of bricks and decorated with tiles; iron stoves were first produced much later, in the sixteenth century.

The open fireplace, with its flue forming part of the outer wall, is a much more recent invention than either the stove or central heating (Roman fashion). It first appeared in Britain in the twelfth century, but it was not common until the sixteenth century.

As a general rule, the most efficient forms of domestic heating have been developed in districts where the winters are severe, or where fuel is scarce. The stove was in general use, in the nineteenth century, throughout Russia, Scandinavia, Germany, and the northern states of America and China. In these countries the open fireplace was a luxury possessed by the few, and used only for supplementary heating on special occasions. In the southern states of Europe and of North America, any form of house heating was a luxury because the number of cold days in a year hardly justified the expense of building fireplaces in living rooms; here the open fireplace was favoured by those who could afford it.

Winter in the British Isles is as long as in the temperate continental regions, but although sunshine is less frequent, the air is warmer and more humid. Perhaps in the circumstances it is natural that open fires remained popular for so long. Their cheerful blaze and radiant heat are a fair substitute for the sun. They do not, however, make a whole room habitable during cold spells.

The establishment of Clean Air Zones required replacement of open coal fires by smokeless sources of heat. The relative cheapness of these fuels and convenience and swiftness of operation contributed much to the successful transition, changing markedly the pattern of domestic heating in the United Kingdom. Although these fires are all able to supply hot water (by heating a back boiler) they only provide a source of heat locally. Nowadays there is a great demand for background heating—to maintain a temperature in the region of 20°C throughout the home. Central heating systems can be fuelled by gas, electricity, solid fuel or oil. Of these four fuels the first two produce little or no pollution locally—pollutants from domestic sources are related only to combustion and hence a major portion of this chapter is devoted to these combustive sources.

The use of oil generally has increased rapidly. However, the large price rises of the early seventies, together with the fuel crises caused by the Arab embargo of 1973-4 and the Iranian crisis of 1979—not forgetting the oil shortfall expected by the end of the century—have cast doubts on the long-term investment in oil. It now seems likely that, as far as combustion of fossil fuels is concerned, an energy strategy based largely on coal will be employed by countries such as the U.K. and the U.S.A. where considerable reserves exist. Concern about rising costs of domestic energy sources has led to an increased implementation (often encouraged by governmental grants) of methods of insulation in order to save energy. Re-use of heat is also of interest; and these topics will be mentioned at the end of this chapter.

Choosing a Domestic Heating System

Few people have an absolutely free choice either in buying their house or in selecting the heating system for it, but the intrinsic value of any dwelling in temperate latitudes depends greatly on its warmth in winter. We shall now consider some of the heating appliances available in Britain and some of their main features. One very important feature is the standard of heating produced by a given rate of fuel consumption; but this depends on the size and shape of the room, the materials of which the walls, floor, and ceiling are made, the temperatures of adjoining rooms, and so on. If variations due to these and similar causes are avoided by confining attention to a particular room, it still depends on the skill of the householder in managing the appliance, the wastage of fuel during periods when the room is overheated, and the time needed for the appliance and the room to warm up.

By the efficiency in ordinary use of an appliance, in the following paragraphs, will be meant the fraction of the potential heat in the fuel effectively used; i.e. not wasted by overheating of rooms, by other imperfections of average management, or as heat and combustible materials escaping up the chimney.

Other important features to be considered when choosing an appliance are its appearrance and the amount and method of cleaning. The cost of the appliance and of installing it, as well as its probable life, should be taken into consideration together with the cost of the fuel it will consume. Many of these points differ among different makes of the same kind of appliance, but only the main features of each type of appliance will be considered here.

Solid fuel

Domestic heating by combustion of solid fuel can be accomplished using an open fire, a modern roomheater or an independent boiler. The first two supply heat locally but can also be used with a back boiler to provide hot water and central heating.

The *open fire* has an extensive history. The cheapest fuel today is Housecoal but if the house is in a Smoke Control Area a smokeless fuel must be substituted since the volatile fraction of coal produces smoke when combusted. Open fires may be simple inset fires, underfloor draught fires, fan-assisted fires or free-standing fires and are up to 60 per cent efficient in running a central heating system. (This can be increased by use of a throat restrictor in the flue).

Until 1920 the design of fireplaces adhered very closely to tradition, although the physicist Count Rumford pointed out a number of valuable improvements in 1796, and although more efficient designs were demonstrated at the Smoke Abatement Exhibitions in London and Manchester in 1882, the Victorian fireplace, with its register and vertical fireback, had an efficiency in ordinary use of less than 15 per cent. After 1920 there was a change in design, and the fireback was made to slope forward in order to reflect into

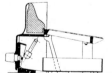

Fig. 51. Fulham grate

the room heat which would otherwise have been radiated up the chimney. There was another big improvement after World War II and Fig. 51 illustrates the Fulham grate which was the first of the type now known as "improved" open fires, whose average efficiency in ordinary use was about 25 per cent. There are now over sixty different open fires approved by the Domestic Solid Fuel Appliance Approval Scheme (see Fig. 52)—available in six sizes: 350, 400, 450, 500, 550 and 600 mm. Ash cans need only be emptied once or twice a week with these fires.

Domestic Heat Services

Fig. 52. Baxi Burnall Underfloor Draught Open Fire (Baxi Heating, Bamber Bridge, Preston, Lancs.)

Roomheaters are a more efficient way of burning solid fuel. They provide heat by radiation and convection whilst retaining the visual aspect of an open fire behind a glass door (in some types this can be opened). With a back boiler, efficiencies of 70 per cent can be achieved. Roomheaters can be controlled by thermostats and may have time controls. Solid fuel can also be supplied by gravity feed in a type of roomheater, which incorporates a hopper (see Fig. 53). Recently introduced, the Prince 76 is able to burn selected

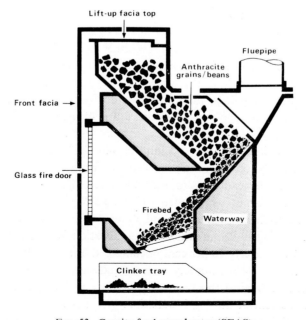

Fig. 53. Gravity feed roomheater (SFAS)

housecoal to the requirements of the Clean Air Act. It is an openable roomheater which can run up to five radiators and provide hot water. The smoke produced is burnt in a second combustion chamber so that there is no residue to pass into the flue (see Fig. 54).

Fig. 54. Prince 76 Roomheater (SFAS)

Central heating

A more elaborate method of house warming than any of the methods considered above, central heating is used in all the temperate regions of the world. The aim is to transfer as much heat as possible from the burning fuel either to the air itself or to a circulatory system of water or steam. The circulating medium normally passes through two forms of "heat-exchanger", first in the boiler where the circulating medium takes in heat, and then through radiators or panels, suitably distributed in the building, where the circulating medium gives out heat to the air and the fabric of the building. Heat is delivered mainly by convection, and the term "radiator" is misleading.

The boiler is often situated directly behind the fire (a "back boiler"). Alternatively the fire may provide only local heating and the boiler for the central heating system may be located elsewhere. Independent solid fuel boilers (see Fig. 55) can achieve efficiencies of 65–75 per cent with outputs ranging from 13 to 35 kW.

Indeed thermal efficiencies between 60 and 80 per cent are achieved by most types of domestic central heating boiler in which the combustion is automatically controlled. The fuel saving may be nearly 50 per cent in comparison with traditional methods with a similar standard of comfort, notably at week-ends and in holiday periods when a dwelling is in full use. Against this economy must be set the capital cost (interest and amortization) of the central heating installation. When the chosen fuel is oil, a fairly large oil storage tank is needed. Solid fuels, too, will need to be stored in some quantity.

Domestic Heat Services

Fig. 55. The "wrap-round" Baxi Radiator Output Boiler (R.O.B.2)
(Baxi Heating, Bamber Bridge, Preston, Lancs.)

Practically any fuel may be used for central heating. Where oil and gas are cheap, they are particularly suitable, because their rate of burning is easy controlled with a thermostat and the labour of maintenance is small.

In 1965 about $1\frac{1}{2}$ million homes had some form of central heating, one-sixth of them using oil as a fuel. In the following 10 years, oil prices began to rise and more and more people installing central heating systems chose gas as a fuel. Consumption of fuel reached a peak in the early seventies and has since been severely moderated by the increasing costs.

District heating (DH) schemes utilize heat from a single central source to heat a whole estate. Economic and other pressures seem to have resulted in procrastination and lack of publicity for such schemes in the United Kingdom; although district heating is widely used in Germany, Scandinavia, Eastern Europe and the USSR. In London, the Pimlico area was at one time served by waste heat from Battersea Power Station (now partially closed) and in Reykjavik hot water is piped from hot springs nearly 20 km away from the city. The virtues of combined heat and power schemes (CHP) have been recently extolled although the present (1978) governmental view is that CHP in Britain is still uneconomic despite the reduction in waste-heat output to the hydrosphere (effectively doubling the efficiency of steam-turbine power stations). It seems highly likely that increasing prices of oil and gas (and decreasing availability) will eventually lead to proliferation of such schemes in the U.K. In future years we may all be encouraged and indeed find it economically beneficial to live adjacent to our place of work rather than removed from it.

Gas and electric fires

Gas and electric fires have the advantage of immediate response and are often preferred, especially when only intermittent heating is required—as when both parents go out to work. Initially ease of installation into existing fireplaces assisted in this swing away from coal. The pollution associated with North Sea gas is minimal; electricity produces no pollution locally although fossil-fuel power stations are responsible for large emissions of pollutants such as water vapour and sulphur dioxide and have notoriously low efficiencies.

Thermal storage electric heating

Because of the "base-load" problem in generation, discussed in Chapter 6, many electricity distributors can offer "night tariff" rates substantially less than the normal cost. Thus it may be advantageous to use an appliance in which heat is developed electrically at night, and stored until the following day. In a "night-storage" heater a well-insulated object of high heat capacity, is used as a heat store and is fitted with a resistance-wire heater of several kilowatts, the heater being switched on principally at night.

Thermal insulation

Many British householders have found that their standard of domestic heating has become much higher than formerly, and that instead of having small "comfort zones" round individual fires, the living area is comfortably warm all the time. An inevitable

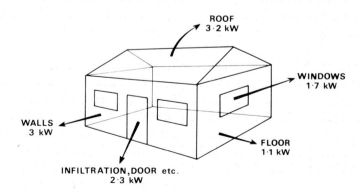

Fig. 56. Typical heat losses from a poorly insulated bungalow

result of the whole house being warm is that heat losses by conduction through walls, windows and the roof become noticeable (Fig. 56). For example it has been shown that about 10 per cent of the fuel needed to heat a typical house is required because of heat lost through the roof. However, this loss can be substantially reduced, in old houses as well as new, by introducing thermal insulation into the roof. The material may be fibre glass, slag wool, or insulating granules placed between the beams over the upper floor ceilings. In some cases aluminium foil is more convenient. The cold water tank in the roof should have lagging over it, but no insulation underneath. With increasing fuel costs, loft insulation is gaining both popularity and importance. (A recent Parliamentary Bill (1979) is designed to encourage this further by giving grants for loft insulation.) Planners of the long-term energy policy of a country would do well to consider wide-scale implementation of insulation and draught exclusion as a prime method of reducing the total fuel consumption of the country. Of slightly less importance are the heat losses through walls and windows, although the larger the area of glass, the larger will be the relative heat loss. This can be overcome to some extent by double or triple glazing and by cavity wall insulation; although at present the pay-back time of about 10 years means that there is little *economic* incentive to implement these modifications.

Solar heating and sensible orientation and internal design of houses (e.g. living quarters on the southern side, work areas on the north) are currently under investigation as a partial (or sometimes full) solution. Granada TV sponsored such developments at the "House of the Future" in Macclesfield (Fig. 57) and in 1979 two council houses in Salford were built to incorporate heat recycling with an estimated energy usage (for heating and hot water) equal to 15 per cent of that required by a normal house.

Fig. 57. "House of the Future", Macclesfield

Hot Water and Cooking

Incidentally to the discussion of the choice of heating appliances, the main types of cooker and water heater which combine several functions have been mentioned. The appliances specially for water heating or cooking include a number which may be easily installed after a house is occupied, with the result that the housewife often has a freer choice of water heaters and cookers than of methods of house heating. There have been such rapid developments since 1919 that in most homes a constant supply of hot water, and a self-regulating cooker, are regarded as essential.

Gas may be used in two ways to provide immediate hot water. The *instantaneous water heater* burns gas only when the water is turned on, heat being transferred to the

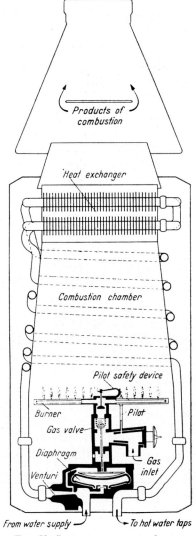

Fig. 58. Instantaneous water-heater

water as it runs through a finned or a coiled pipe (see Fig. 58). The *storage water heater* burns gas whenever the water in a tank cools below a specified temperature, and this happens when hot water is drawn from the tank and replaced by cold water. Gas is not burnt as fast as in the instantaneous heater, and after a whole tank full of water is used there is an interval of perhaps 40 min before the water is fully hot again. The first newly heated water, however, collects in the top of the tank, and small quantities of hot water are available after a few minutes. Both appliances have a heating efficiency of up to 75 per cent if several tens of litres of water are required; but they are much less efficient whenever small quantities are used at a time. The average in general use might be about 50 per cent. For comparison, the efficiency of a gas ring in heating a kettle is about 40 per cent.

Water may be heated by electricity in the same two ways. An electrically heated storage water-heater is illustrated in Fig. 59. The efficiency of the electric appliances is about

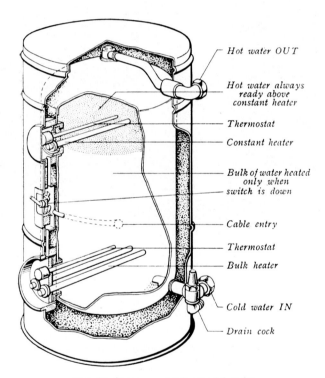

FIG. 59. Storage water-heater (electric)

98 per cent; i.e. practically all the potential heat of the current is transferred to the water, the only losses being due to the finite resistivity of the wiring. It is often simple to convert a hot-water tank into an electric storage heater by adding an immersion heater, but the tank should be thermally insulated with a generous layer of lagging if electricity is not to be wasted.

Gas and electric cookers are too well known to need any description. In the course of trade rivalry between gas and electricity boards, their potentialities have been very well developed, and nowadays you can even select "mix and match" cookers. There would be advantages in having an electric oven with gas rings for heating kettles and saucepans, and perhaps an electric grill and toaster. Solid fuel cookers (see Fig. 60), developed from the old-fashioned kitchen range, are gaining enthusiastic supporters throughout the country. Some of these cookers have hot plates on which a 2 litre kettle can be boiled in 2 min after the overnight banking.

FIG. 60. The Aga CB Cooker, for households up to eight people plus integral water heating. (Courtesy of Aga Ketley)

Coal economy

If an individual householder wishes to choose a heating appliance, a question then arises which may be put as follows: What type of domestic heating appliance should be encouraged in the national interest? To answer this question it is important, besides atmospheric pollution, to consider the economic use of coal. The efficiencies given above refer to the actual fuel used whether primary or derived, and those referring to derived fuels would all be lower if coal were considered as the starting point.

The heat which used to be lost at gasworks in converting coal into coke and gas averaged about 27 per cent, i.e. the average efficiency of conversion was 73 per cent, but some gasworks achieved 80–85 per cent. A figure of 85 per cent may be used, for example, if a national plan is being prepared which would entail the building of a number of new gasworks (as may be necessary when oil and natural gas reserves become severely depleted). Low-temperature coke can also be produced with an efficiency of 85 per cent. The efficiency of generation of electricity in the newest plants is about 34 per cent.

This chapter has treated at length the problems of domestic heating in Great Britain, for a number of reasons. First, the methods now in use are at least as varied as in most other countries. Second, they are rapidly changing in relative importance and this work, while primarily concerned with atmospheric pollution, may have a secondary value in helping changes to be made where they are most needed. Finally, domestic fires in Britain are still responsible for at least some of her atmospheric pollution, in spite of efforts made since World War II, and particularly since the Clean Air Acts 1956 and 1968, to eliminate this pollution. No other country has such a history of pollution from domestic fires, and few countries have had more to gain than Great Britain by the elimination of domestic air pollution. However, in view of the predicted global fuel shortages, governmental and personal decisions of the future will have to be made in the light of resource management and economics as well as air quality.

CHAPTER 10

Atmospheric Pollution

THE country dweller entering a large town can usually detect, by the nose, the eye, or the sensation of the skin, a definite deterioration in the quality of the air. The deterioration is more noticeable in some towns than others, and in any particular town it is more noticeable on some days than others; but even when town air is at its best the country air is better still. It used to be commonly believed that town air is stale by reason of having been breathed by so many people, yet in the most densely populated areas the hourly breath requirements of the people would be satisfied by a layer of air less than 1 cm thick. The only detectable chemical differences between town and country air are differences due to atmospheric pollution.

The composition of air is given in Table 21. Among the major constituents, water vapour is extremely variable, its amount depending strongly on temperature and relative humidity. The concentrations of the major constituents can also be expressed as per cent by volume, and the minor constituents in units that are 10,000 times smaller: parts per million.

TABLE 21. *Approximate composition of the dry atmosphere*

	(i) Major constituents		(ii) Minor constituents		
	$g\ m^{-3}$	% (by volume)		$mg\ m^{-3}$	ppm
Air	1184	100	Neon	14.9	18
Nitrogen	895	75	Helium	0.85	5.2
Oxygen	274	23	Methane	0.79	1.2
Argon	15.2	1.26	Krypton	3.4	1.0
Water vapour	var	0.70	Hydrogen	0.04	0.5
Carbon dioxide	0.57	0.04	Nitrous oxide	0.90	0.5
			Xenon	0.43	0.08

Nearly all the atmospheric pollution in Britain is made by the burning of fuels, although there are contributions from fuel handling, chemical processes, and natural sources. Table 22 shows an estimate by staff of the Warren Spring Laboratory, of pollution produced in 1959, 1967, 1976 by various classes of fuel user. Pollution emitted during the manufacture of gas, coke and pig iron is not included, nor was any study made of the emission of fine ash, grit or the rarer chemicals in fuel.

TABLE 22. *Pollution emitted from fuel in Great Britain, million tonnes per year*
(1959, 1967, 1976)

	1959		1967		1976	
	Smoke	Sulphur dioxide	Smoke	Sulphur dioxide	Smoke	Sulphur dioxide
COAL						
Domestic	1·2	0·9	0·80	0·69	0·33	0·19
Power stations	—	1·3	—	1·90	—	2·12
Railways	0·2	0·3	0·02	0·02	0·00	—
Industrial etc.	0·5	1·5	0·12	0·97	0·04	0·27
Coke ovens	—	0·1	—	0·08	—	0·13
Gas works	—	0·2	—	0·10	—	0·08
COKE, excluding gas works and blast furnaces	0·0	0·4	—	0·28	—	0·14
OIL	0·0					
Diesel and gas oil	—	0·05	—	0·43	—	0·51
Fuel oil, power stations	—	0·2	—	0·44	—	0·67
Fuel oil, other uses	—	0·6	—	1·42	—	0·94
Total	1·9	5·5	0·94	6·33	0·37	5·05

Note: The above rounded data (for 1959) were derived from "Trends in the pollution of the air of Great Britain by smoke and sulphur dioxide 1952–59" by J. D. Carroll, S. R. Craxford, H. E. Newall and M-L. P. M. Weatherley; *Proceedings of the Clean Air Conference,* Harrogate 1960, published by the National Society for Clean Air.

Smoke

Before going at all deeply into detail about atmospheric pollution, it is essential to have a clear understanding of the differences between smoke, ash, sulphur dioxide and other important forms of pollution. Smoke is the term normally applied to the visible products of imperfect combustion. A smoke plume* from a chimney may be long or short, but ultimately the smoke becomes so well mixed with air that it ceases to be visible, except possibly as a bluish haze which obscures distant objects. After such attenuation smoke remains, however, a potential cause of dirt and damage, and it can still be measured with the help of instruments. In this book it is still called smoke when it has become diluted by so much air that it is invisible.

Coal smoke contains a high proportion of carbon, and when viewed in bulk it is nearly black. It also contains tarry hydrocarbons, which add to its sticking powers and its tendency to form sooty deposits in chimneys and elsewhere.

---

* Many chimney plumes today are, however, visible mainly due to their high liquid water content. Smoke plumes are allowed, for very restricted periods, under the controls of the Clean Air Act.

Smoke has the important property that, because of the small size of its particles, it behaves in many ways like a gas and has the same powers of penetration. The average diameter of a smoke particle is about 0·075 μm. From Stokes' Law of the flow of fluids past spherical objects, it can be shown that all particles smaller than about 10 μm are easily supported by the bombardment of air molecules. They do not fall to the ground very rapidly under their own weight, but are swept this way and that by every current of air. So, when air enters a house through windows, ventilating bricks, or even through cracks round the doors, particulates enter too; and some are deposited as dirt on walls and ceilings, curtains and furniture. Air molecules are continuously bombarding the walls and other surfaces; so are the particles, but while the gas molecules bounce off, the particles tend to stick and make the surface dirty.

Smoke also sticks to the outside walls of buildings, for rain will not wash it away, unless the stone is slightly soluble or very smooth. Although on one or two particular buildings small deposits of smoke have been considered rather attractive, smoke has been responsible for the dingy appearance of very many urban buildings. A striking effect occurs on smooth walls of white limestone; in places under projections which protect them partially from the rain, the smoke sticks and produces a half-washed appearance (see Fig. 123).

Quite obviously, smoke does not remain permanently in the atmosphere, and the average time for which a smoke particle remains in suspension has been estimated as one to two days. It has been suggested that many of the particles deposited on land are caught near the edges of objects such as blades of grass and the leaves and twigs of trees, where small eddies may occur. Electrostatic attraction may also play a part in the process.

Smoke is not the only form of atmospheric pollution which consists of extremely small particles. Some of the ash particles which escape from chimneys are so finely divided that they, too, remain suspended in the air. In dry windy weather, fine dust may be blown up from the ground, and also drops of sea spray will evaporate, leaving minute particles of salt which may be blown far inland.

Clean Air legislation has resulted in a large reduction of visible smoke emissions. A decreased number of suspended particulates in the air is noticeable in the increase of visibility. In a large conurbation such as Manchester this is clearly evident (see Fig. 98). A smaller pollutant load scatters les of the son's radiation and brings about an increase in the number of sunshine hours (see Fig. 124). The average size of particulates present in the atmosphere is, however, much smaller than formerly; such small (often microscopic) particles can penetrate deep into the longs. Present concern about particulate pollution is centred largely on those of respirable size ($\lesssim 10$ μm) which are implicated in respiratory diseases such as bronchitis and emphysema (see Chapter 14).

Ash

Ash is the unburnable solid material that is set free when a fuel is burnt. The sparks, which fly upward from a coal fire, are red-hot particles of ash. In a fire or furnace where the fuel is coal or coke much of the ash falls through the fire bars into the ash pit, but an appreciable proportion escapes with the flue gases. Coal contains from 2 to 10 per cent or more of mineral matter, but only a fraction of this is released in a finely divided state when the coal is burnt on a grate, and the ash which is small and light enough to be carried up the flue represents, on a very general average, about 0·3 per cent of the coal. If the coal is finely ground and burnt as pulverized fuel, however, most of the ash passes into the flues.

Assuming there is no dust arrestment, the maximum size of particles carried up any particular flue depends on the velocity of the flue gases or, more strictly, on the velocity at the point in the flue-system where opward flow is slowest. Industrial flue gases have velocities up to about $12\,\mathrm{m\,s^{-1}}$, fast enough to carry up particles as large as 0·2 cm (or 2000 μm) in diameter. Most industrial chimneys are less than 60 m high, so, in the absence of freak wind currents, the largest particles of ash emitted would remain in the air for less than 5 sec. Because of their short time in the air, large particles of ash have never formed a substantial part of the "suspended impurity" in the air, but they have contributed appreciably to the pollution deposited on the ground.

In the chimneys of open fires, the flue gases seldom have a velocity exceeding $2\,\mathrm{m\,s^{-1}}$, and the particles escaping are all less than about 75 μm. All larger particles remain in the grate or the ash pit, and this is why only a relatively small amount of ash is emitted in the domestic use of coal. It has been variously estimated that in the 1950s 0·5–1·5 million tonnes of ash were emitted each year from British chimneys. Only about one-fifth of the atmospheric ash came from domestic chimneys; the remainder, including nearly all the largest and most unpleasant of the particles, came from a variety of industrial sources.

From certain installations where pulverized coal is burned as much as half the grit emitted is combustible matter, and the particles recovered from the flues and electrostatic precipitators constitute a fuel of considerable calorific value, although its practical uses are limited.

The amount of grit blown from ash heaps and coal dumps depends very much upon the wind, the sizes of particles present, and the dampness of the heaps. Clouds of particles may also be set free when coal, ash, or other materials are being loaded or unloaded. As a general rule, however, grit from sources such as these is not a large contribution to the pollution of a town, though it may be a nuisance within a few hundred metres of its place of origin.

Sulphur dioxide

Sulphur dioxide is formed in considerable quantity when coal, coke or certain fuel oils are burnt. Though it is not so chemically active as sulphur trioxide, hydrochloric acid, and the fluorine compounds which are also liberated during the combustion of coal, it is emitted in much greater quantity and is thus capable of doing more harm.

The average burnable sulphur content of British coal, and also coke, is about 1·3 per cent, with extreme values of about 0·9 and 4·0 per cent. When the fuel is burnt, only a small amount of the sulphur remains with the ashes, and the rest is released as sulphur dioxide in the flue gases. Of this a small amount is usually retained in the flues and, if no waste gas removal is undertaken, the rest is emitted to the atmosphere.

From a consideration of atomic weights, it is seen that a molecule of sulphur dioxide weighs twice as much as an atom of sulphur. So for each tonne of sulphur in the fuel there are nearly 2 tonnes of sulphur dioxide in the flue gases, and it is estimated that an average of nearly 3 tonnes of sulphur dioxide is emitted from every hundred tonnes of coal or coke burnt. Over 5 million tonnes of sulphur dioxide are emitted each year from chimneys in Britain.

Sulphur dioxide is a gas, with a choking smell. In the open air of industrial towns, sulphur dioxide can sometimes reach high enough concentrations to be faintly smelt (the odour threshold is approximately 600 µg m^{-3}). Like smoke, sulphur dioxide penetrates indoors, where it tarnishes metals and makes fabrics tender and easily torn after washing.

Sulphur dioxide is soluble in water, and it is particularly liable to attack paint, metals, stone work, and slates when water is present. When rain water wets the surfaces and interstices of a building, and sulphur dioxide is present in the air, a very dilute solution of sulphur dioxide is formed which becomes more concentrated later as the water begins to evaporate. Further, in the presence of air this solution is soon oxidized to form sulphuric acid which is a very reactive substance. It appears probable that the corrosive action of sulphur dioxide is chiefly due to the formation of sulphuric acid.

Some sulphur dioxide is removed from the air by solution in cloud droplets (rainout) falling rain (washout), and in surface water, but only a small percentage of the sulphur dioxide emitted to the atmosphere is brought down with rain. The remainder is either absorbed by wet surfaces (e.g. buildings, vegetation) on land or remains in the atmosphere to be blown away from our shores. Its ultimate fate must be to reach the ground by one of the mechanisms outlined above in a different part of the globe. A large proportion of the sulphur dioxide emitted from Great Britain is thought to be responsible for acid rain in Scandinavia. The proportions of sulphur dioxide escaping from its country of origin to that being removed near its original emission site depends largely upon (i) the height and location of its emission (i.e. whether or not it escapes the influences of the urban heat island, the local topography etc.) and (ii) the chimney height and the prevailing meteorological conditions (i.e. whether there is sufficient turbulence to dilute the plume sufficiently before it can affect ground level concentrations).

Carbon monoxide and carbon dioxide

Concentrations of carbon monoxide from 10 to 70 mg m^3 are common in busy streets; concentrations of 120 mg m^3 or more are considered dangerous. Most of it is due to the incomplete combustion of petrol in internal combustion engines; properly adjusted diesel engines burn a "lean" mixture and produce very little carbon monoxide. Although the incomplete combustion of coal and coke also produces carbon monoxide, it is doubtful whether enough of this could reach street level to be a danger to health.

Carbon monoxide is a colourless, odourless, tasteless gas which is toxic at sufficiently high concentrations because of its higher affinity for the haemoglobin in the blood than oxygen. It is present in vehicular exhaust (petrol engines only) and, to a larger degree, in cigarette smoke. It has a threshold limit value (TLV) of 50 mg m^{-3} (50 ppm)—see Chapter 14 for full discussion of threshold limit values.

Carbon monoxide can be readily oxidized to carbon dioxide—a product of complete combustion. Carbon dioxide is an odourless non-toxic gas, usually regarded as a desirable end product (together with water) of combustion of hydrocarbons.

Carbon dioxide is a normal constituent of the atmosphere, necessary to vegetable life, but an unpublished observation from the British Oxygen Company's works in Greenwich, where carbon dioxide has to be removed from the air before it is liquefied, indicates that during the London smogs and fogs of the fifties the concentration of carbon dioxide was well over twice the normal amount. A sufficiently high concentration, of about 10 or 100 times normal, would accelerate human breathing and enhance the effects of poisonous gases. It also enhances photosynthesis by plants, which take up the excess carbon dioxide together with any other noxious gases present.

Recent concern over increasing levels of carbon dioxide is based on the climatological effect of the gas. Carbon dioxide absorbs and "traps" terrestrial radiation which then heats up the Earth. This so-called "greenhouse effect" could increase the global surface temperature by 1 or 2°C if the carbon dioxide level is doubled (if our use of the combustion process continues to rise at the present rate it has been suggested that this will occur before the middle of the next century).

Nitrogen oxides (NO_x)

There are three major oxides of nitrogen, known collectively as NO_x. Nitrous oxide, N_2O, is only produced naturally. Both nitric oxide, NO, and nitrogen dioxide, NO_2, are emitted anthropogenically. Both result from oxidation of nitrogen present in fossil fuels and, more commonly, from nitrogen in the air used in the combustion process. Vehicles emit large concentrations of nitric oxide which is rapidly oxidized in the atmosphere to nitrogen dioxide—a reaction heavily implicated in the formation of photochemical smog (see below). Nitrogen dioxide is a brown gas with an odour threshold of about 200 μg m^{-3}. Emissions of NO_x for 1970 are given in Table 23.

TABLE 23. NO_x emissions in the U.K. in 1970 (units: 10^3 tonnes)

Transport	270
Domestic	68
Commercial	524
Power stations	600
Total	1462

Lead; chlorine and fluorine compounds

Lead compounds, especially those added to petrol as anti-knock agents, are emitted to the atmosphere and are, at present, a great cause for concern, since lead is an accumulative poison and is more likely to be ingested by children. The health effects of lead are discussed in Chapter 14.

Among its numerous mineral constituents, coal contains up to 0·7 per cent of chlorine and up to 0·01 per cent of fluorine. When coal is burnt, both these elements can escape to the atmosphere in the gaseous forms HCl (hydrochloric acid), HF (hydrogen fluoride) and SiF_4 (silicon tetrafluoride). All these compounds, particularly those containing fluorine, can do harm to men, animals and materials.

Pollution from Petroleum Products

In Los Angeles, where petroleum products (including natural gas) are used almost exclusively for heating and power, there are special conditions of climate and topography which aggravate the effects of pollution. What was once a health resort is in danger of failing to support its now enormous population of 5 millions in reasonable cleanliness and comfort. In 1954 a conference was called by the Southern California Air Pollution Foundation to discuss "Vehicle combustion products and other emissions". The proceedings of this conference gave a first insight both into the problems of Los Angeles and into the risks associated with pollution from sources other than coal.

The Los Angeles Basin is an area of about 1000 sq. km with the Pacific Ocean to the west and hills on the other three sides. On about 260 days in the year light or zero winds and a temperature inversion below 300 m impede natural ventilation. In such conditions air at ground level is changed perhaps only once per day. A smog develops which reduces visibility (an especial danger for air traffic), and produces eye irritation and crop damage.

The pollutants discussed above are all (with the exception of nitrogen dioxide) termed *primary pollutants* since they are all emitted directly as a result of a combustion process. Atmospheric chemistry may then produce *secondary pollutants* from this pollutant "soup", e.g. the oxidation of nitric oxide to nitrogen dioxide as discussed above. In Los Angeles and other large conurbations where oxides of nitrogen and unburnt hydrocarbons are emitted, strong sunshine can energize reactions producing complex secondary pollutants which form a *photochemical smog*.

Although the full chemistry of such smogs has not yet been fully determined, there has been much progress since the first investigations in the 1950s. In contrast to a reducing sulphurous smog, a photochemical smog is a strong oxidant resulting from the production, amongst other pollutants, of ozone which damages plants, cracks rubber and is an eye irritant. It is now thought that smog formation follows the chemical history outlined below.

Nitric oxide is emitted in large quantities by vehicle exhausts. In the atmosphere this is rapidly oxidized to nitrogen dioxide. The important role of solar radiation is now seen, because nitrogen dioxide absorbs wavelengths of 0·38–0·60 μm and is dissociated to nitric oxide and atomic oxygen. Atomic oxygen is quickly "mopped up" in a three-body reaction to form ozone.

A second cycle is initiated when hydrocarbons are introduced (as in dense traffic). The hydrocarbons compete with the molecular oxygen for the atomic oxygen. The result is a long series of complex reactions yielding a vast range of organic molecules, many of which are toxic and/or irritants, e.g. acrolein, formaldehyde, peroxyacetyl nitrate (PAN) and possibly the carcinogen, peroxybenzoyl nitrate (PBN).

Largely confined to the eastern seaboard of America, the phenomenon of photochemical smog has, however, been recently observed in Britain during the hot summers of 1975 and 1976 when large increases in concentration of ozone and nitrogen oxides were observed in some of our major cities.

Odours

It is unlikely that odorous emissions have greatly increased over the past two decades; yet over this period concern about their effects and effective control has increased considerably. This is most probably a result of the decreased level of other pollutants such as smoke and sulphur dioxide. In many cases odorous emissions are simply a nuisance causing little harm in terms of human health.

Identification of objectionable odours (malodours) is difficult since this involves a large degree of subjectivity. For instance, the distinctive smell of hops from some breweries may be enjoyable to some (because of its associations!) or found to be bitter and pervading. Any odour can be described in terms not only of its *quality* but also its *intensity*. At a certain concentration (a different value for different people and different odours) the smell is detectable. This is the *odour detection threshold*. A larger concentration is needed for identification (the *odour recognition threshold*). Although threshold values vary between observers, a "typical" threshold value can be identified—usually by a panel of trained observers (e.g. ammonia, 47 ppm; hydrogen sulphide, 0·000 47 ppm; sulphur dioxide, 0·47 ppm). Not all odours with low thresholds are malodours, e.g. perfumes—artificial musk, 0·000 007 ppm).

Odour identification is still most efficiently undertaken using the human nose; although there is presently much research into both the application of standard pollution techniques (e.g. gas chromatography) and the development of new methods (e.g. olfactometers) in odour assessment. Odour control is often undertaken using absorption techniques (adsorption on activated carbon), by afterburners or by neutralization or masking (e.g. deodorants which contain a "pleasant" odour stronger than the malodour).

The problems of odours emitted from specific industries in the United Kingdom have been recently investigated by a Working Party of the Warren Spring Laboratory (1974/5).

Radioactive air pollutants

The atmosphere transports and deposits radioactive isotopes in the same manner as their less harmful chemical counterparts. However, the problems are not limited to movement from source to sink but also encompass decay times and long-term effects. The major component of all radioactivity in the atmosphere is from natural sources (Fig. 61).

The development of nuclear power production has brought with it a new kind of pollution problem, for there is a danger that the atmosphere, and also land and water, may locally become contaminated with radioactive waste materials, which must be removed from the plants in which the "fuel" is prepared or consumed; and radioactive materials cannot be destroyed—they can only decay naturally in the course of time. Our body cells can withstand only a fairly small amount of radioactivity, and we cannot safely breathe, eat, or remain close to substances that are badly contaminated with radioactive wastes. A medium radioactive dosage will produce temporary changes in the blood. A stronger dose, received in a short interval, may produce nausea and other symptoms of radiation sickness. Four or five times this dose might be fatal.

Too much radiation over a period has been shown to produce genetic effects. No worker is allowed in Britain to receive an annual dose of radiation of more than 75 rem (to the skin) or to be subject to a background greater than $\frac{1}{8}$ rem. The level for the general populace is set at $\frac{1}{2}$ rem although it should be noted that the International Commission for Radiological Protection (ICRP) recommendations based on complicated empirical equations have recently replaced more straightforward guidelines.

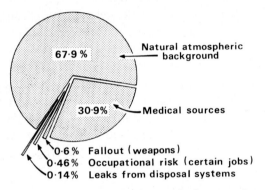

Fig. 61. Origin of radioactive isotopes in the atmosphere

Safety has top priority in nuclear power plants: workers undergo daily checks both for their own safety and to reduce the risk of removal (on clothes for instance) of radioactive dust from the confines of the nuclear plant. The risks associated with nuclear power are perhaps put in perspective when one considers that we receive much more radiation from X-rays than any other source (except natural background)—see Fig. 61.

Radioactive dust or gases emitted from stacks are severely restricted and storage of spent fuel is undertaken with the utmost stringency. New methods, such as vitrification, are being developed to reduce risks even further.

Unfortunately small-scale accidents have occurred. In 1956 an accident in one of the experimental reactors at Windscale, Cumbria, resulted in the emission of about 20,000 curies of radio-iodine through a 120 m high stack. This was more than the entire world stock of radioactive materials in 1938. Some of the radio-iodine was deposited in grass and eaten by cows. As a result the cows' milk was contaminated and it had to be destroyed for many weeks.

In 1979 the development of a gas bubble within the reactor at Three Mile Island in Harrisburg in Pennsylvania resulted in the nearest approach yet to a major disaster in the Nuclear Power Industry and has given much food for thought for both protagonists and antagonists of nuclear energy. Indeed emissions resulted in the neighbourhood being subject to a radiation level a factor of 15 times greater than normal background levels.

The explosion of nuclear bombs either in the theatre of war or in atmospheric testing leads to radioactive fallout when it can be a major hazard. However, nuclear arms limitation treaties and test bans have been largely responsible for the decreased importance of fallout in terms of an air pollutant (see Table 24).

TABLE 24. *Concentration of strontium-90 in rain, U.K. (in picocuries per litre)*

1972	2·62
1973	1·19
1974	2·68
1975	1·92
1976	0·94

Pollution from Other Sources

Paint particles and solvent odours can be very troublesome particularly in districts which include motor-car manufacture. Air filtration and water washing systems may be included in paint-spraying booths with ultimate discharge of the effluent gases from chimneys at least as high as 40 m. Even then, very finely divided droplets may escape and the odour of solvents remains. At the present time the best hope with the large-scale paint-spraying plants associated with car-body manufacturers is to ensure, as far as possible, that their location is such, either within the factory site or outside it, that there is ample open space in the immediate vicinity to allow harmless fall-out of droplets and diffusion of odours.

In the case of smaller manufacturers the wet washer and air filter can be satisfactory provided that the water is not recirculated beyond the point of efficiency and that the plant is regularly maintained.

Gases from chemical works

In the very early days of the Alkali Industry, sodium carbonate was made by the Leblanc process, involving the strong heating of a mixture of sodium chloride and sulphuric acid; and the hydrochloric acid gases were allowed to escape into the atmosphere. The damage to vegetation and property was so great that the hydrochloric acid had to be recovered by solution in water, in coke-packed towers down which water was sprayed. The solution of hydrochloric acid was soon found to be commercially useful, and an important by product industry developed. To ensure the adequate removal of hydrochloric acid the first Alkali, etc., Works Regulation Act was passed in 1863, specifying a condensation of 95 per cent, and an additional Act in 1874 limited the escape to one-fifth of a grain of muriatic acid (hydrochloric acid) per cubic foot of flue gas.

Without the incentive of this Act it is possible that through less thorough methods of recovering hydrochloric acid there might be a threat to public health. In its present form the Act protects the public against all harmful substances liable to be emitted from industrial sources other than from fuel burning (see Chapter 16).

Among other gases from chemical works which cause atmospheric pollution are oxides of sulphur, sulphuretted hydrogen, carbon bisulphide, nitrogen oxides, chlorine and hydrogen fluoride. Concentrations and other conditions are so different that not only each polluting gas but each process has had to be considered individually, in order to develop a satisfactory method of reducing the polluting gas and, if possible, producing a useful product.

Sulphur dioxide is emitted, for instance, from the roasting of sulphide ores. It may be recovered and converted into sulphuric acid by the contact process. On the other hand, from the large-scale manufacture of sulphuric acid, emissions of sulphur dioxide and sulphur trioxide can best be reduced by scrubbers of high efficiency.

Sulphuretted hydrogen is evolved in numerous processes, such as the carbonization of coal, the distillation of tar and ammonia liquor, the distillation of petroleum, and the manufacture of artificial silk and paper by the viscose process. This poisonous gas can be recovered by absorption by iron oxide placed in layers in large boxes or "purifiers". The iron sulphide collected in this way by the carbonization industries used to be used for the production of sulphuric acid, yielding (in the 1950s) annually in Great Britain some 270,000 tonnes of acid out of a total of 1,800,000 tonnes.

Carbon bisulphide in the vapours from certain waterproofing processes can be absorbed by activated carbon, and subsequently recovered by steaming the carbon. In low concentrations both sulphuretted hydrogen and carbon bisulphide are difficult to absorb, and recovery processes become impracticable if large volumes of gases are to be treated. In such cases the only practical safeguard may be by combustion, forming sulphur dioxide which, though harmful, is less noxious than either sulphuretted hydrogen or carbon bisulphide.

Oxides of nitrogen, which are liberated in many nitrogen processes, and the chamber sulphuric acid process, are dangerous gases. They are most effectively absorbed by first oxidizing the nitric oxide to nitrogen peroxide and then scrubbing the gases in towers irrigated with water.

Escapes of chlorine are usually only accidental as the gas is readily absorbed either in soda solution or milk of lime. Emissions of fluorine and its compounds from superphosphate works and hydrofluoric acid works are adequately controlled in Great Britain by the provisions of the Alkali, etc., Works Regulation Act. Recently there have been emissions of fluorine from factories manufacturing aluminium, near which concentrations of 0·02–0·10 mg of fluorine per cubic metre were measured. Fluorine compounds are also emitted, in small amounts, at brickworks and pottery kilns, and from the calcination of ironstones, if fluorine is a constituent of the raw materials used.

Although on a first consideration the gases liable to be emitted from chemical works are very alarming, there are two redeeming features. First, most gases which have a rapis, chemical action in the human being also react rapidly with water or other substanced and so they are usually easy to remove from the effluent gases. Second, reactive chemicals

Atmospheric Pollution

are nearly always useful, and there is an incentive to the manufacturer to recover them. Thus the pollution of the atmosphere by chemical works, although it remains a very serious problem, is not so great an evil as might have been expected.

Burning spoilbanks

The spontaneous ignition of stored coal was referred to in Chapter 3. In colliery districts there is the serious risk that spoilbanks will ignite spontaneously and go on burning for very long periods. Colliery spoil contains coal and pyrites which produce smoke and sulphur dioxide, at ground level, in offensive and damaging concentrations. Spoil fires can now be easily detected by infrared aerial scans.

The ideal treatment is to remove the spoilbank completely. To "back-stow" debris underground into worked out pit galleries is excellent, but it is slow and expensive, and can deal with only about half the total. Tipping into ravines or making strictly controlled embankments on low-lying land are possible methods, but no layer should be

FIG. 62. Spoilbank before removal

FIG. 63. Spoilbank during removal

deeper than 3 m. The top 15 cm of soil should be removed beforehand from the site and returned to cover the tip, and fine dust separated by screening from house refuse may be added; both help to exclude air from the combustible material beneath.

In preparing for new spoil banks, all the above measures should be considered. Where possible, combustible matter should be removed; the spoil should be crushed, and the spoil bank systematically levelled and consolidated by mechanical means. The form of the tip should be a low embankment—conical tips are to be avoided as the most liable to ignite (see Figs. 62–64).

FIG. 64. Spoilbank after removal

Incineration of refuse

The burning of commercial and industrial wastes in yards or on factory sites is objectionable and troublesome, and is subject to the Clean Air Act, 1956 in Britain. The easiest way to avoid the offence, and often the cheapest, is for the local authority to collect the rubbish, or to accept its delivery to their own refuse tip or incinerator. They are then required to make a "trade refuse" charge under the Public Health Act, 1936, Section 7.

Incineration can be undertaken on either a small or a large scale. Municipal incineration is becoming increasingly common, but not necessarily sited in the traditional locations away from the urban areas (see Fig. 65). The furnaces are designed to operate with a heterogeneous fuel—ranging from paper to animal carcasses; domestic to selected industrial rubbish. The system may employ a fixed or moving grate and some incinerators contain a grateless furnace with a horizontal rotating drum or fluidized cyclone. Odours and vermin can sometimes be a problem with stored rubbish. Fires sometimes occur because of spontaneous ignition of the rubbish. Stack emissions are largely water since spray mechanisms and dust-control devices are usually employed; although in many plants little direct continuous monitoring of the chemical constituents of the waste gases is carried out.

Fig. 65. Plume from Salford incinerator

The offensive trades

This is the title given in the Public Health Act, 1936, to trades concerned with the sterilization and drying of animal, fish and vegetable refuse to produce valuable feeding stuffs. The gases evolved have a very objectionable smell, as they contain aliphatic amines and sulphuretted hydrogen. They may either be passed with the air of combustion through a furnace, where they are completely oxidized by burning, or they may be treated with chlorine or hypochlorite.

Particles

In industry there is the risk of polluting the air, not only by gases, but by solids. Examples are blast furnaces, cement kilns, smelting works, and coke ovens. Solids are removed by devices similar to those adopted for the removal of solids in the combustion of coal. The most common are the cyclone, the wet scrubber, and the electrostatic precipitator (see Chapter 15). The last is particularly useful for recovering valuable metallic dusts.

In this book no attempt is made to discuss what is described as the industrial dust hazard. Nearly all cases of pneumoconiosis, including asbestosis, silicosis in coal mines and other industries, and byssinosis in the cotton and flax industries, occur as a result of breathing dust indoors or in confined spaces. The simplest way of avoiding a dusty atmosphere is often by drawing the dust, with ventilating hoods, away from the place where it is generated. The dusty air from the ventilating hoods may be filtered and recirculated, to conserve the warmth which it possesses, although this procedure is risky, since few filters completely remove those particles, smaller than 1 or 2 μm, which do most damage to the lungs. It is generally safest to discharge the filtered air to the atmosphere where it is usually so rapidly diluted that little harm is done. Complaints of harm or discomfort from the dust so emitted to the atmosphere from ventilating systems are rare.

CHAPTER 11

Measurement of Atmospheric Pollution

THE earlier chapters have shown how potential atmospheric pollution can be traced and studied with considerable accuracy while near the fuel bed and on its way up the chimney; but when it emerges from the top of a chimney its individuality begins to be lost and it rapidly becomes unidentifiable among the pollution from other chimneys and other districts. For this reason accurate experiments with pollution in the open air can only be made with great difficulty. Although some useful inferences are possible from controlled experiments and theory nearly all that is known of open-air pollution is the result of observation and measurement, taking conditions as they are, rather than of experiments in which pollution is specially produced.

For the scientific study of pollution on a full scale the same general methods are used as in studying weather. Systematic observations are made at selected places, over many years, and the accumulated data are examined month by month and year by year. Since 1914 a large number of measurements of atmospheric pollution have been made in Great Britain, mostly by local authorities, government departments and firms. Many of the results have been published annually by the Stationery Office, on behalf first of the Air Ministry, then the Department of Scientific and Industrial Research and later of the Department of Trade and Industry, where the tasks were undertaken of coordinating and collating observations, as well as of devising methods of measurement and investigating the properties of atmospheric pollution.

In two important respects, atmospheric pollution is different from weather and climate. The climate of any particular place changes only very slowly, but the atmospheric pollution may alter considerably within a few years; observations of pollution are particularly important where industrial or residential development is taking place, or where special efforts are being made to improve the purity of the air. Secondly, the amount of pollution is often quite different in adjacent districts, whereas the rainfall, barometric pressure, temperature and other aspects of weather are often nearly the same over large areas. It is therefore important to study the distribution of pollution, i.e. its variation from district to district. One of the first serious surveys of air pollution was undertaken in Leicester in 1937–39. Many, both large and small scale, surveys have been carried out since in addition to the continuous monitoring of the National Survey co-ordinated by the Warren Spring Laboratory.

The three principal types of open-air pollution from chimneys are smoke, particulates and sulphur dioxide. Where they constitute a nuisance they can often be seen or smelt, and simple observations can be made without instruments. For the purposes of records and comparison of observations, however, standard methods of measurement are necessary, and it is now general practice in the British Isles to adopt the methods recommended by the Warren Spring Laboratory of the Department of Trade and Industry. They are still evolving, but there is a great deal of continuity with past observations under the guidance of a long series of public-spirited Atmospheric Pollution Research Committees. (The National Society for Clean Air has long been a pioneer in the field.) The general aim has been to devise methods which are simple and labour-saving. From most points of view it is preferable to have a large number of approximate measurements of pollution rather than a few extremely accurate ones. Indeed, pollution varies so much with time and place that great accuracy would be needed only if an extremely large number of observations were being made.

Measurement of Smoke

Prior to the Clean Air Act, 1956 observations of smoke emitted by chimneys were undertaken by the Public Health Inspector (now known as the Environmental Health Officer) in order to determine whether the emissions were a nuisance or whether the emission of black smoke contravened any by-laws under the Public Health Act, 1936. No definition was given of the phrase "black smoke" and individual assessments varied. Under the Clean Air Act regulations have been made giving "permitted periods" during which dark smoke and black smoke may be emitted. The smoke densities are themselves related to the Ringelmann scale of shades, dark smoke being smoke of Ringelmann 2 or denser and black smoke of Ringelmann 4 or denser (see Fig. 66).

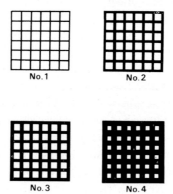

FIG. 66. Ringelmann chart

The observer stations himself in some convenient place to observe the top of the chimney, as in Fig. 67, taking care not to confuse the shade number of the smoke by taking observations into the sun or along the line of the smoke. The time is recorded and subsequent emissions of smoke are noted depending on their colour and duration.

Observations such as the above are of great practical value in enabling local authorities to limit the emission of smoke, but they hardly deserve to be described as measurements. Attempts have frequently been made to assist the eye and judgement of the observer by instruments and scales. To ensure a correct assessment of its shade the smoke is compared directly with the Ringelmann Chart placed at the correct distance, or assistance is gained by comparison with a miniature reproduction of the full-size chart, known as a micro-Ringelmann or with the aid of a comparison telescope such as the "Telesmoke", which is a pocket telescope containing an eyepiece graticule ruled like the Ringelmann Chart (see inset of Fig. 67). A full-scale observation may extend up to 8 hours and to aid this work some authorities are experimenting with automatic cameras, time-lapse photography and even radar or lidar.

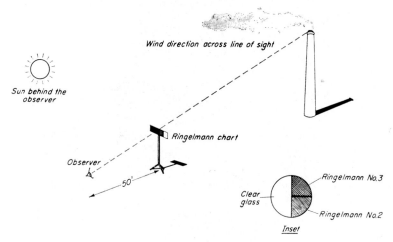

FIG. 67. Method of smoke observation, using Ringelmann Chart. Field of view of Telesmoke. *(Courtesy of W. Ottway & Co.)* The image of the smoke close to the head of the chimney stack is kept in the clear part of the field of view, best viewing distance about 100 metres

The general concentration of smoke in the air, as distinct from the weight of smoke emitted from any particular chimney, can be measured in a number of ways. If air is drawn through filter paper, practically all its solid matter is removed and remains embedded in the filter paper, mostly near the surface. After about 2 m^3 of air have been drawn through a circle of white filter paper 2 cm in diameter, the solid matter on the paper is usually visible as a grey stain, sometimes tinged with brown.

The material collected in this way consists mainly of small particles, because large particles fall out of the air in a relatively short time and have little chance to be drawn into the apparatus. The material may be roughly analysed by finding the proportions which are combustible (like smoke and soot) or soluble in carbon bisulphide (like the tarry hydrocarbons). Much of the knowledge which has so far been accumulated about atmospheric smoke has been deduced from measurements of "smoke" stains. It would have been a far harder task to obtain the same knowledge by the more laborious method of analysis.

The amount of matter on a very dark smoke stain can be weighed on a sensitive chemical balance, but the technical difficulties of weighing are considerable. This is more feasible when using a high volume sampler, as in the U.S.A., which samples over 20 times more air in a given time than the standard British (Warren Spring) sampling procedure. It has been found that the degree of blackness (i.e. the reduction in optical reflection coefficient) corresponds reasonably well with the weight of matter which has been collected; and for many purposes it is convenient to estimate the weight by comparing the stained filter paper with a standard set of shades, each of which corresponds under average conditions to a known weight of the material.

Fig. 68. The photoelectric smoke reader.
(Courtesy of Evans Electroselenium Ltd.)

For readings of greater accuracy, the optical reflection coefficient of the stained filter paper may be measured with a photoelectric cell. Perfectly clean filter paper, of the variety normally used, has a reflection coefficient of 85 per cent. For small weights of smoke, the weight is directly proportional to the amount by which the reflection coefficient falls short of 85 per cent. The photoelectric smoke reader shown in Fig. 68 operates on this principle. When greater weights of smoke are collected, successive increments of smoke make less and less difference to the reflection coefficient.

For measuring the concentration of smoke in the open air filter paper or glass-fibre paper is used in three of the four methods now to be described.

Smoke filter

The standard apparatus for measuring the daily average concentration of smoke is illustrated in Fig. 69. (It is shown combined with a bubbler for the volumetric estimation of sulphur dioxide, to be described later.) Air is drawn continuously through a filter paper by a small suction pump. The filter paper is changed daily, and the area on which the smoke is collected is a circle of diameter 2·5 or 5 cm, according to the situation, time of year, and volume of air sampled each day. The volume of air is measured with a suitable meter, placed between the filter and the pump. (The order of apparatus—inlet, collection device, measure, pump—is adhered to for all pollutant sampling). The main apparatus is under cover, usually inside a room, but the air which is being tested is drawn from out of doors through a glass tube which projects about 1 metre from the wall of the building and terminates in a small inverted funnel, to keep out quickly falling particles and drops of rain. Little smoke is lost by deposition in the tube, provided that the flow of air is non-turbulent.

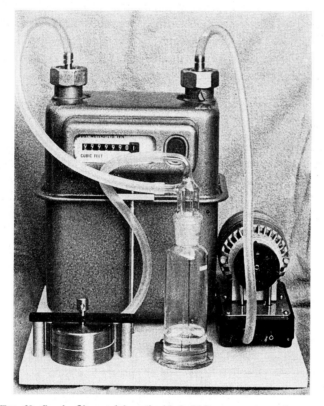

Fig. 69. Smoke filter, sulphur dioxide bubbler, gas meter and pump.
(Courtesy of AGL.)

The whole apparatus has been standardized and the Warren Spring Laboratory of the Department of Trade and Industry has issued guidance notes on the measurement of atmospheric pollution by means of the volumetric apparatus.

For determining the average concentration of smoke collected by a filter, a photoelectric reflectometer is used (Fig. 68). This measures the reflection of light by the smoke stain as a percentage of the reflection of light from the unstained paper. A conversion table obtainable from the Warren Spring laboratory converts percentages, given the volume of air, to *milligrammes* of smoke *per cubic metre* (mg m^{-3}). This unit is now used for the measurement of all forms of suspended impurity—smoke, sulphur dioxide, and other gases. As this unit is rather large, a unit of one-thousandth the size, namely *microgrammes per cubic metre* (μg m^{-3}), is more often used instead.

Once the smoke filter has been installed, regular daily readings may be taken with the expenditure of less than 5 min/day. Although the results are of considerable usefulness and interest, too great an accuracy should not be claimed for them. Errors may be caused (1) by inaccuracy of the gas meter, particularly as it is working at a slight under-pressure, (2) in matching a smoke stain with the standard scale of shades, (3) by applying the calibration of the scale of shades, or of the smoke meter, to other atmospheres than that of London in winter. As a general rule, it would be unsafe to attribute an accuracy within 15 per cent to any single determination of the concentration of smoke in the air, unless special precautions were taken. The ratio of two determinations taken in similar circumstances could probably be relied on to within about 5 per cent, provided that each determination involved the measurement of ten or more smoke stains.

Self-changing smoke filters

Smoke filters in which there is automatically inserted a clean piece of filter paper at intervals of about an hour are useful, particularly for studying the daily cycle of smoke. The most historical of these instruments is the *automatic filter* designed by Dr. J. S. Owens in 1918. It is shown diagrammatically in Fig. 70, omitting the clock and weighted string which moves on the filter paper after each siphoning.

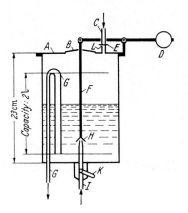

FIG. 70. Automatic filter

Modern electrically operated self-changing filters produce larger smoke stains on a long strip of filter paper (see Fig. 71). Glass fibre paper is more sensitive, though more delicate, than standard filter paper made from vegetable fibres. A specimen of the exposed filter paper is shown in Fig. 72.

Fig. 71. Self-changing smoke filter

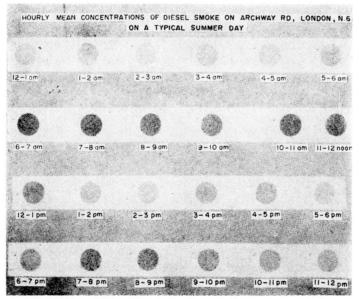

Fig. 72. Record of diesel smoke with the self-changing smoke filter

Portable smoke filters

It is possible to measure the concentration of smoke at any particular place in 5–10 min. With a suitable pump, 2 or more litres of air are drawn through filter paper, leaving a smoke stain 0·32 cm in diameter. The pump may be operated by hand, by foot, or by electric motor driven from a car battery. The volume of air may be measured with a gas meter, or in the case of a hand or foot pump, the volume pumped at each stroke may be measured in a separate test.

Weighable smoke filter

It is usually necessary to draw air through a filter for several days at least, before a quantity of smoke suitable for weighing can be obtained. Filter paper has been used, but only with difficulty, because it tends to become choked, and because it retains a quantity of moisture, often weighing much more than the smoke, which varies according to the humidity of the air. Asbestos fibre is superior to paper as it is not hygroscopic, and it can be readily made into a filter pad through which only a very small proportion of smoke particles can pass. Filter paper of woven glass fibre is available of a quality which is more efficient than either vegetable paper or asbestos as a retainer of smoke particles; even this material should be brought into equilibrium with air of a standard humidity before it is weighed.

Measurement of Ash and Other Deposited Pollution

If the air contains significant amounts of ash and grit, the impurities may be collected for examination and analysis by various fast-flow air sampling instruments. These include impingement devices in which the air is blown on to a small reservoir of water; and impaction devices in which the air is blown on to a dry surface.

A much simpler rapid survey can be undertaken under dry conditions by exposing a sheet of glass, covered with a thin film of vaseline, for several hours. However, since such a rapid survey will only produce interesting deposits in particularly polluted areas, care must be taken to consider spatial and temporal aspects of sampling, both for site selection and in the analysis of the results.

Deposit gauge

In 1912 a number of local authorities in Britain agreed to unify their procedure in measuring deposited atmospheric pollution. The deposit gauge which they adopted has been reduced a little in size, and undergone other slight modifications, but remains in essentials the same. It is illustrated in Fig. 73. and now the subject of British Standard 1747–1951. For best results, the position of the deposit gauge must be chosen to represent the surrounding districts as fairly as possible. Buildings and trees create upward and downward currents of wind, causing abnormal deposits, so they should be at least twice their height from the deposit gauge.

The solid and liquid material, including rain, which falls within the 30 cm diameter circle of the glass collecting bowl, passes down the connecting pipe, of rubber and stainless steel, into the glass bottle. Material which falls on the outside of the collecting bowl and on the galvanized iron stand is prevented from entering the bottle by the inverted stainless-steel funnel. The galvanized wire screen gives some protection to the bowl and serves to prevent birds from settling on its edge.

On the first day of the calendar month the deposit gauge is set up with a perfectly clean collecting bowl, connecting pipe, and bottle. The bottle contains 10 ml of a 0·02 per cent solution of copper sulphate, or about 1 g of parachlorphenol, to prevent the growth of algae which would alter the chemical nature of the materials collected. On the first day of the following month the gauge is examined, and any identifiable "foreign matter" such as leaves or insects is removed. A burette brush is passed downwards through the connecting pipe and the outlet of the collecting bowl is blocked with a bung, through which the wire of the brush penetrates. About half a litre of the collected water is poured into the bowl and the deposited matter adhering to the inside of the collecting bowl, which may sometimes amount to more than half the total, is dislodged with a rubber squeegee. The bung is removed, and the water with the solid matter in suspension is allowed to run into the collecting bottle, while the inside of the connecting pipe is cleaned with the brush.

FIG. 73. Deposit gauge (height overall, 1·5 m)

The bottle which has been exposed for one month is handed, with particulars of site, period of exposure, and the exact area presented by the collecting bowl, to a qualified analytical chemist. The following are the steps of the analysis, technical details being omitted.

(1) The volume of water is measured.
(2) The pH value of the water is measured with Universal Indicator.
(3) The collected material is separated, by decanting and filtration, into insoluble matter and filtrate.
(4) The total insoluble matter is dried and weighed.

(5) The weight of "tarry matter", i.e. insoluble matter which is soluble in carbon bisulphide, is found.
(6) The insoluble matter which remains after extraction with carbon bisulphide is ignited by heating to 800°C in a crucible, and the residual ash is weighed.
(7) The weight of combustible matter, other than tar, is found by difference.
(8) A measured fraction of the total filtrate is evaporated to dryness, dried at 100°C and weighed.
(9) The weight of calcium in a measured fraction of the total filtrate is found.
(10) The weight of chloride is found, similarly.
(11) The weight of sulphate is found, similarly, after oxidation of any sulphites which may have been present.

The analyst records the results on a special form, and they are converted to convenient units, as the example in Table 25 shows. Corrections are made for the small volume of water and the small weight of sulphate which were originally put into the bottle to prevent the growth of algae.

TABLE 25. *Material collected by deposit gauge in one month in 1958; example of analysis*

1. Volume of water		litres	
Rainfall		2·73	38 mm/month
2. pH value			7·1
		grammes	mg m^{-2} day^{-1}
4. Total undissolved matter		0·1873	86
5. "Tarry matter"		0·0043	2
6. Ash		0·1098	50
7. Combustible matter, other than tar		0·0732	34
8. Total dissolved matter		0·0846	39
9. Calcium		0·0096	4
10. Chloride		0·0173	8
11. Sulphate		0·0415	19
12. Total solids (total of undissolved and dissolved matter)		0·2719	125

Notes: The numbers 1–11 in the table correspond to the numbers of the steps of analysis, in the text above.

Table 26 reflects the changes (not large) in deposits over the last 20 years (cf. Table 25). There is, as expected, a difference between urban and rural sites, although summer values are higher than winter values—an opposite relationship in comparison with smoke and sulphur dioxide data.

Results of a single month's observations with a deposit gauge can only be relied on within a standard deviation of about 20 per cent, because of errors of sampling. When a deposit gauge is exposed for a number of consecutive months at the same site, it is found that the successive monthly estimates of each type of pollution vary by more than 20 per cent. The standard deviation is at most places about 40 per cent of the average: this means that about one reading in three is different from the average by more than 40 per

TABLE 26. *Distribution of annual summer and winter mean deposits, 1973–4 at urban sites, and at country sites monitoring particular sources percentage of sites with deposition exceeding valve in left-hand column,*

Percentage of sites with deposition ($mg\ m^{-2}\ day^{-1}$)	1973–4		Summer 1973		Winter 1973–4	
	Urban	Country with sources	Urban	Country with sources	Urban	Country with sources
50	87	77	87	81	76	59
100	40	25	48	32	30	14
150	18	9	22	15	13	5
200	9	4	11	9	6	2
250	4	2	5	4	5	1
300	3	2	3	2	3	0
Median ($mg\ m^{-2}\ day^{-1}$)	88	69	97	82	74	56

Data source: Warren Spring Laboratory Reports.

cent. This large variation has been found to be due partly to changes in the rate of emission from local chimneys, but mostly to fluctuations in meteorological factors such as monthly rainfall, and direction and speed of wind.

The effect of the large variance is to render necessary a long period of observations before significant conclusions can be drawn about the rate of emission of pollution. In practice, the most useful conclusions have been drawn from 5-year groups of 60 monthly observations, though significant changes have sometimes been detected within a few months of their occurrence.

In the post-1956 era in Britain, the role of deposit gauges has diminished in importance. The National Survey of atmospheric pollution has been designed to study the concentrations in the air of smoke and sulphur dioxide, because of their importance both for medical studies and in connection with clean-air legislation. Deposited matter is regarded as a more local problem, which can be measured and dealt with on the initiative of local authorities. In many cases the figure for insoluble deposit is the only one that is generally useful, and the analysis may then be limited to items 1 and 4 of Table 25.

Directional dust gauges have also been developed to attempt to differentiate between dust originating from different directions at low angles and impinging on a vertical surface. Analysis is undertaken monthly as with the standard deposit gauge.

Rapid surveys of deposited matter

Long-term trends tend to ignore fluctuations in the weather. Different information can be obtained by considering pollutant levels under specific meteorological conditions. Rapid surveys to this end can be carried out by sampling particulates using petri dishes. These are round glass dishes, with short vertical sides, provided with lids similar in shape but slightly larger; often about 9 cm in diameter. Besides a supply of petri dishes, the only apparatus required is a good balance, two small watch glasses, and a small camel-hair brush.

At least nine, but normally more, sites for dishes are chosen every 400 m ($\frac{1}{4}$ mile) within the area to be surveyed, care being taken to select sites, preferably about 3 m (10 ft) above the ground, where interference by the public or by animals is unlikely. The distance from every object (such as trees and buildings) should be at least twice the height of the object above the level of the dish.

When anticyclonic conditions prevail, and there appears to be reasonable prospect of enjoying at least 24, and preferably 48 hr of dry, calm, weather the dishes are polished clean and distributed. The lid may be inverted and placed beside each dish. After 24 or 48 hr, the dishes are collected and labelled, each being inverted and placed inside its lid, so that none of the solid matter is lost. If the dishes are distributed and collected in the same order the period of exposure is approximately the same for each one.

Weather conditions during the exposure of the petri dishes should be carefully watched, for the smallest amount of rain or a gust of wind at more than about 5 m/sec (12 miles/hr) will spoil the experiment. At intervals observation should be made of any change in weather or of any special source of pollution. The wind direction should be noted by observing smoke from a tall chimney; weather cocks, flags near the ground, and even clouds are not always reliable indicators. The wind velocity should be estimated roughly according to the standard Beaufort scale of the Meteorological Office, the relevant part of which is summarized in Table 27.

TABLE 27. *Beaufort Wind Force and equivalent velocity*

	Force (arbitrary units)	Velocity mi/hr	Velocity m/sec
Calm; smoke rises vertically	0	0	0
Direction shown by smoke drift only, no pronounced sensation of air movement	1	1–3	1
Wind felt on face; leaves rustle	2	4–7	2–3
Leaves and small twigs in constant motion; light flag extends	3	8–12	4–5
Dust and loose paper raised (experiment invalidated by dust blown into and out of the dishes)	4	13–18	6–8

In the laboratory the contents of each petri dish and lid are brushed on to a watch glass with a dry brush, and weighed in the watch glass to a high accuracy.

Besides the weight of material collected in each dish, it is necessary to know the period of exposure and the area presented by the dish. For the degree of accuracy required, which is about 5 per cent, it is usually sufficient to take the period and the area as the same for all dishes. It is then possible to construct a map showing contours (isopleths) of equal deposit (see, for example, Fig. 74). It shows at a glance where deposits were heaviest and where they were least during the period of exposure, and may readily be compared with maps showing the built-up area or the location of industry in the district.

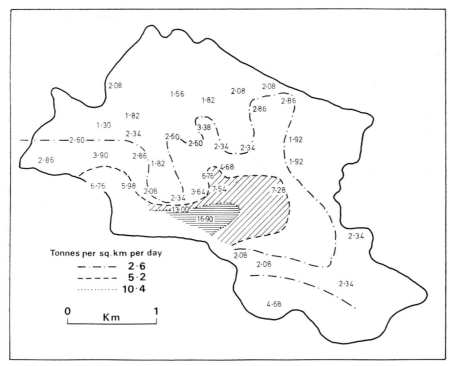

Fig. 74. Petri-dish survey of deposited matter in Bilston, 14–16 Dec. 1943

The effect of a chimney from which excessive quantities of solid particles are emitted is to produce a region of heavy deposit within a few hundred metres downwind, which would be instantly noticeable on a map such as Fig. 74. Petri dishes may therefore be used to help in accumulating evidence against a chimney which is suspected of causing a nuisance, but for this purpose it is best to expose dishes at intervals of about 200 m in the affected area.

Compared with the deposit gauge, petri dishes have the advantages of numbers, speed in producing useful results, low initial cost and cost of operation. Their main disadvantages are the exacting nature of their demands on personnel, the uncertainty that suitable weather will prevail and that, since observations can only be made in particular weather conditions, petri dishes cannot be used for observing average rates of deposit. Though samples taken with petri dishes cannot be analysed as thoroughly as with the deposit gauge, the proportion of combustible matter and ash can be found; prior to analysis, a general idea of the nature and origin of the particles can be got by microscopic examination as described in a later paragraph.

Measurement of Sulphur Dioxide

As part of the analysis of deposited matter collected each month by the standard deposit gauge, the weight of sulphate in solution can be determined. Most of this sulphate is formed by the solution and oxidation of sulphur dioxide which has been emitted from chimneys as a gas. It can be imagined in a similar way that sulphur dioxide must be collected by the roofs and walls of buildings, and be oxidized in the presence of water to sulphuric acid and sulphates. The weight of sulphates collected in a deposit gauge is therefore an index of the amount of sulphuric acid, derived from sulphur dioxide, which damages the external surfaces of buildings, and also corrodes metals and other materials exposed in the open.

Sulphur dioxide is, however, harmful in other ways which are not satisfactorily measured by the deposit gauge. It is particularly important to know the concentration of sulphur dioxide in the air breathed by men, animals, and plants. The absolute determination of sulphur dioxide in air which may also contain alkalis and other active materials is a laborious task for the chemist. However, the following method is simple, economical in labour, and accurate enough for most ordinary purposes.

Volumetric estimation of sulphur dioxide

Air from outside a building is drawn at a steady rate of about 2 m³/day through a bubbler, the volume of air being measured with a gas meter. The bubbler contains a dilute solution of hydrogen peroxide, about 3 cm in depth, 25–40 ml in volume. All the sulphur dioxide is effectively removed from the air bubbles by this small quantity of liquid, and sulphuric acid is formed:

$$H_2O_2 + SO_2 = H_2SO_4.$$

At the beginning of the test, the solution of hydrogen peroxide is brought to a pH value of 4·5, by the addition of small amounts of acid or alkali, using B.D.H. "4.5" indicator which is pink at pH values less than 4·0, grey at 4·5, and blue at values greater than 5·0. At the end of the test, which usually lasts 24 hr, the pH value is less than 4·5 because of the sulphuric acid which has been formed in the solution, and the amount of alkali of known strength is measured which is required to bring the pH value back to 4·5. In practice, the alkali is sodium hydroxide or, better, because of its superior keeping qualities, borax of strength N/250 (i.e. containing 1/250 gramme-equivalent of the reagent in a litre of water). For N/250 alkali, the average concentration of sulphur dioxide during the period of the test is calculated by the formula

$$\text{Concentration of } SO_2 \ (\mu g \ m^{-3}) = 128 \cdot 0 \times \frac{\text{ml of N/250 alkali}}{m^3 \text{ of air}}.$$

It is usual for the bubbler of the volumetric sulphur dioxide apparatus to be preceded by a smoke filter, as in Fig. 69. The average daily concentrations of smoke and sulphur dioxide can be conveniently measured with the combined apparatus, and it has been found that a negligible amount of sulphur dioxide is absorbed from the air as it passes through the filter paper. Many ordinary materials, however, including soft glass, brass, and rubber, absorb sulphur dioxide quickly enough to affect the volumetric determination appreciably. It is therefore important to use polythene or hard-glass tubing, and to apply a coat of vaseline to the internal surfaces of rubber and brass.

Provided that there is no absorption by the inlet tubing, all the sulphur dioxide entering the inlet funnel is caught by the bubbler. This can be demonstrated by having a second bubbler in series with the first. The main limitations of the method are due to the absorption in the bubbler of other gases which are acid or alkaline.

Carbon dioxide from the air is moderately soluble in water, forming a weak acid (carbonic acid). All difficulties from this source are removed, however, by the choice of 4·5 as the initial and final pH value. A saturated solution of carbon dioxide in pure water has a pH value of 4·5, and dilute solutions having this pH value are not appreciably changed in pH by the addition or removal of carbon dioxide. In addition to sulphur dioxide and carbon dioxide, air usually contains traces of ammonia, hydrochloric acid, and sulphur trioxide. It is doubtful whether sulphur trioxide (which becomes sulphuric acid in solution) is trapped by the bubbler, since it occurs in the atmosphere as sulphuric acid mist which will be caught by the filter, but in any case the amount of sulphur trioxide is small, perhaps a hundredth of the sulphur dioxide in the air.

There is usually more of the alkali ammonia in the air than there is hydrochloric acid. If both these gases are efficiently trapped in the bubbler, as seems likely, the ammonia will neutralize all the hydrochloric acid and in addition a little of the sulphuric acid formed from the sulphur dioxide. The concentration of ammonia in the air is generally not more than 0·01 parts per million, sufficient in the absence of hydrochloric acid to neutralize 0·005 parts per million of sulphur dioxide. Therefore, although estimates of sulphur dioxide are often expressed to the nearest 0·001 part per million, the last figure is doubtful. Another possible way of looking at the matter is to regard the estimates as determinations of the total acidity of the air. To affect plants, animals and materials, sulphur dioxide must as a rule go into solution in the surface moisture which is always present; but ammonia, being more soluble, is likely to be taken up first. Hence the sulphur dioxide in solution will probably have to neutralize this ammonia before the rest of it can attack other substances or be oxidized to sulphuric acid. There is therefore some justification for the view that the total acidity of the atmospheric gases is of more practical significance than the total sulphur dioxide. In towns the two interpretations differ so slightly that there is little risk of confusion.

Portable instruments for sulphur dioxide

For volumetric sampling studies of the distribution of sulphur dioxide, e.g. in smog conditions or in the vicinity of a strongly emitting source, the ideal sampling time is generally a few hours. It must be long enough for samples taken at slightly different starting and finishing times to be regarded as simultaneous, and yet short enough for conditions of emission and weather to be assumed as unchanging. A simple form of portable instrument was developed at the Warren Spring laboratory. It is driven from a 6-V battery, and the volume is measured by a counter recording the number of strokes of its oil-immersed piston pump.

Instantaneous measurements of sulphur dioxide present in chimney plumes can now be made by a ground-based observer by using ultraviolet television sensing. The equipment, which is microprocessor-controlled, was developed at NASA's Langley Research Center in Virginia.

Fig. 75. "Eight-port" volumetric sampling apparatus for SO$_2$ (Courtesy of AGL.)

Automatic monitoring

The bubbler described above must be changed every 24 hours. To avoid this constant attention, especially over weekends, the "8 port" volumetric sampling apparatus shown in Fig. 75 is in general use by co-operating bodies of the National Survey. A timing mechanism redirects the inflow to seven of the eight bubblers in turn (the eighth is a reserve). Titration of the hydrogen peroxide solution is identical: but can now be undertaken weekly.

Use of the "8 port" results in mean daily concentrations of sulphur dioxide. Continuous monitoring can also be used to detect peaks (and troughs) within that 24-hour period. The apparatus (see, for example, Fig. 76) is usually costly and needs regular servicing. Many are based on measuring the acidity in a hydrogen peroxide solution in terms of its electrical conductivity; although a flame photometric method can be used in which hydrogen is burnt in the air stream—sulphur compounds produce a luminescence which can be utilized to estimate sulphur dioxide concentrations above 25 μg m^{-3}.

A continuous trace from a conductimetric Wösthoff analyser is shown in Fig. 77.

Sulphur dioxide by the lead dioxide instrument

In 1932 at the Building Research Station a method of estimating atmospheric sulphur dioxide was devised in which the gas was absorbed by a solid surface in much the same way as it is absorbed by building stone. The surface was prepared from lead dioxide which absorbs sulphur dioxide according to the reaction

$$PbO_2 + SO_2 = PbSO_4$$

Although methods of measurement which are based on this reaction give an indication of the action of sulphur dioxide on surfaces generally, they have not yielded as much

FIG. 76. Meloy continuous monitor for sulphur compounds.
(Courtesy of Techmation, Edgware, Middx.)

information, when used in research projects, as had been hoped. An account of the so-called Lead Peroxide Instrument has been given in the earlier editions of this book. Some of the ways in which its results can be misleading are discussed in a paper by S. R. Craxford, D. W. Slimming and E. T. Wilkins to the Annual Conference of the National Society for Clean Air in 1960.

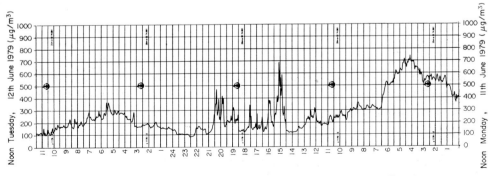

FIG. 77. Continuous trace of SO_2 for June 11/12 1979 from Wösthoff analyser

Pollution Roses

A useful method of data presentation (especially for gaseous pollutants such as sulphur dioxide) is the *pollution rose*. Over a period of, say 1 month or 1 year, the observed mean concentration of pollution associated with each of the ranges of wind speed is represented by a bar pointing in the wind direction. The length of the bar is proportional to the concentration. When the wind was in the SW (Figure 78) the highest concentrations

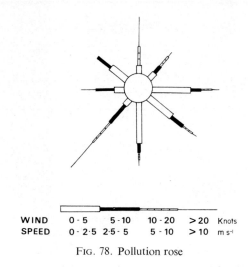

WIND SPEED	0 - 5	5 - 10	10 - 20	> 20	Knots
	0 - 2·5	2·5 - 5	5 - 10	> 10	m s⁻¹

FIG. 78. Pollution rose

were associated with the highest wind speed range (>10 m s^{-1}) and were 50 per cent greater than those for either the 5–10 m s^{-1} or the 2·5–5 m s^{-1} range. However, for E, SE or S winds the largest concentrations occurred for low wind speed (0–2·5 m s^{-1}).

Microscopic Examination of Grit

The microscopic appearance of gritty deposits of atmospheric pollution is often a clue to their origin (see Fig. 79). Samples for the microscope may be collected by scraping some of the deposit into a glass tube, but a better way is by exposing petri dishes for a suitable period in dry calm weather. Petri-dish samples are more free from contamination and can be related to the weather at the time of their collection. At about the same time as outdoor samples are taken, it is advisable to collect control samples of the grit inside the flues of any furnaces suspected of causing the pollution. If possible the control samples should be taken from inside the chimney stack at the base, or from some other position where there is an accumulation of particles just too heavy to be lifted out of the stack.

Representative portions of the samples, including the controls, should be mounted on microscope slides, with cover slips lifted clear of the largest particles by mounting rings of about $\frac{1}{2}$ mm thickness. The specimens should be examined under a microscope with strong top illumination and fairly strong sub-stage illumination, of such magnification and resolving power as to show clearly any particle of a size between about 5 and 100 μm (0·005–0·1 mm). The size range of particles in each specimen should be noted. A description follows of the distinguishing features and manner of formation of various particles likely to be met.

Coke spheres are formed when small particles of coal, swept up from the fuel bed by the gases of the combustion chamber, become semi-molten and rounded, but escape before being burnt. Sometimes the spheres contain blow holes made by volatile matter escaping from within them just before they harden. Not only coal particles, but oil drop-

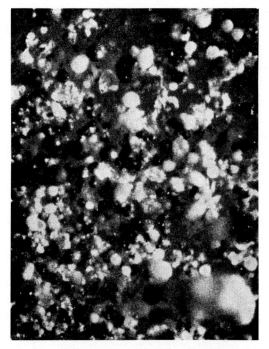

FIG. 79. Fly ash from a boiler fired by pulverized fuel

lets also, may produce coke spheres if they are imperfectly burnt; coke spheres from oil burning are characterized by their roundness, uniform size, and numerous blow holes. When the chimney of a stoker-fired or oil-burning installation continuously emits black smoke, the "smoke" invariably contains a high proportion of coke spheres.

Coke spheres with fly-ash markers occur if smaller particles of fly ash come in contact with the coke particles while they are semi-molten. Fly-ash markers are a sure sign that the particles come from the burning of coal.

Unrounded or partly rounded coke particles are formed when partly burnt coal falls through the grate into the ash pit of a furnace.

Coke fragments from coke ovens are very dense and highly reflective. Since they are usually broken from bigger pieces they have sharp edges and corners.

Ash spheres occur if the furnace is hot enough to melt the particles of fly ash in the combustion chamber. Most large industrial furnaces but practically no domestic fires attain the requisite temperatures. There is an important distinction between ash spheres from pulverized-fuel installations and from those with an ordinary fuel bed supported on a grate. During the combustion of a fuel bed the different minerals in the fuel are well mixed, and the ash is a mixture of the aluminium silicates of iron and calcium. As a result the ash spheres from a fuel bed are coloured, yellow if mainly of calcium and red if mainly iron.

When coal is pulverized and burnt suspended in air, the ash spheres are plain, for there is no opportunity for the minerals of different particles to mix and the shale and pyrites are kept apart. The shale forms white or colourless spheres and the pyrites form black spheres of magnetic iron oxide. Brownish or greyish spheres can be formed if a pulverized coal burner is operated with too little air for combustion, but they originate more frequently when fine coal is burnt on a travelling grate and particles are lifted off the grate before coking sticks them down.

Microscopic examination of suspended matter

Suspended particles are mostly too small to be examined for shape and colour, and many of them are too small to be graded according to size or even to be seen in an ordinary microscope. If they are collected by allowing them to settle on to a microscope slide, the sample contains a grossly unfair proportion of larger particles. It is even less satisfactory to collect the particles by drawing air through a filter paper and putting the filter paper under a microscope, because they form aggregates in which the individual particles are indistinguishable. A number of instruments have been devised, however, for taking microscope samples of the particles suspended in the air of factories or mines.

FIG. 80. Thermal precipitator

The Thermal Precipitator, illustrated in Fig. 80, was described in 1935 by H. L. Green and H. H. Watson. It depended upon the principle, noted in the nineteenth century by Aitken, that when a hot body is placed in a dusty or smoky atmosphere, it can be seen with a lens to be surrounded by a dust-free space. The boundary of this dust-free space is quite definite, and dust is prevented from penetrating within it by the bombardment of fast-moving air molecules. The thickness of the dust-free space depends on the temperature, and is about 0·1 mm if the hot body is 80°C warmer than the air. Therefore if dusty or smoky air is drawn slowly through a sufficiently narrow slot, at the centre of which is a hot wire, the air will go past the wire but the particles will be unable to pass and will be deposited on the sides of the slot.

The *cascade impactor* uses the inertia of particulates to remove them from an air flow. Air is forced round a series of right-angle bends (see Fig. 81). At each bend particles in a

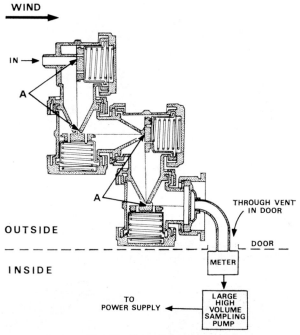

Fig. 81. Cascade impactor

selected size range are unable to follow the air flow and are collected on cover slips for later microscopic inspection. Using a high volume pump, sufficient particulates for analysis can be collected in $\frac{1}{2}$–1 hr.

The microscope is not a perfect instrument for examining small particles, because only particles larger in diameter than about 0·25 μm are visible in it. With the electron microscope, however, smaller particles, down to about 0·025 μm, can be "seen". An electron microphotograph of smoke particles is shown in Fig. 82. When records containing about 100,000 particles were examined, it was found that half the individual smoke particles were smaller than about 0·075 μm; but half the weight of smoke was in particles larger than about 0·51 μm. Some of the properties of smoke depend on the area it can cover;

half the surface or sectional area of the smoke was in particles larger than about 0·35 μm. There is a definite tendency, as Fig. 82 shows, for smoke particles to stick together in chains, perhaps one micron long.

Respirable particulates are those small enough to be inspired and are thus of immediate concern in environmental health. Many monitors have been developed to estimate the concentrations of dust particles of sizes below about 10 μm in diameter. A recent development by T.S.I. is the Respirable Mass Monitor (see Fig. 83) which measures particulates greater than 0·01 μm up to a maximum selected by the operator. An inertial collector

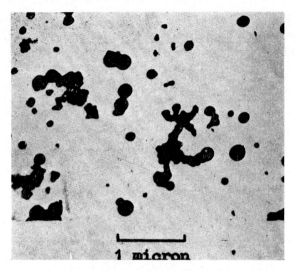

FIG. 82. Electron microphotograph of smoke particles

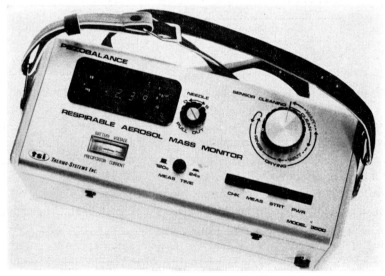

FIG. 83. TSI Respirable Aerosol Mass Monitor Model 3500.
(Courtesy of BIRAL(TSI), Portishead, Bristol.)

removes particles greater than say 5 or 10 μm so that the total mass fraction of particulates can be estimated by collecting those below this cut-off limit by electrostatic precipitation ono a piezo-electric crystal.

Other pollutants

Although smoke and sulphur dioxide are the major pollutants to be continuously monitored, other chemicals in the atmosphere may be of more immediate, local concern.

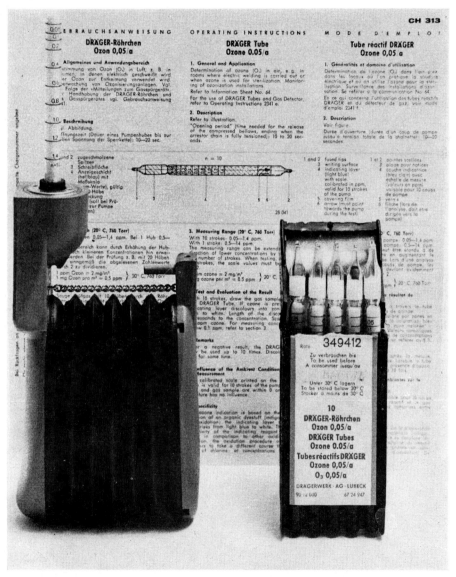

FIG. 84. Draeger tubes and hand-operated sampling pump.
(Courtesy of Messrs. Draeger Safety.)

Qualitative and quantitative assessment of trace pollutants can be undertaken simply and cheaply by use of Draeger tubes (see Fig. 84). More detailed analyses may employ the following methods: gas chromatography, infrared spectroscopy, flame spectrometry, atomic absorption spectroscopy using samples acquired either by "grab sampling" or continuous sampling and absorption by a suitable medium. (In the laboratory the pollutant is usually released from the absorbant by rapid heating.) Rain-water samples can be chemically analysed to give information on the pollutants which were present in the atmosphere before being removed by either rainout or washout.

Measurement of Daylight

There are many instruments for recording daylight, from the Campbell–Stokes hours-of-sunshine recorder to spectrophotometers and pyrheliometers. The only instruments which can be conveniently applied to the study of atmospheric pollution are those which are simple enough for several to be used in different parts of a town, and at the same time sensitive enough to record the difference between the daylight received in a clear and a polluted district.

Instruments which record the total amount of light received as a number or a counter are in use in certain research establishments. The light is made to fall on a photoelectric cell which passes a current proportional to the amount of light it receives. A simple solarimeter such as this is useful in determining either instantaneous or continuous values of incident solar radiation.

Use of measurements of atmospheric pollution

Like all routine observations, measurements of atmospheric pollution are liable to be forgotten after they have been once examined, arranged in a table or a graph, and perhaps published. If they have been well planned, they are worth more care than this, for with proper treatment they will yield information about both the local or general trends of pollution and the effects of particular conditions of weather. Some of these tests are simple, but others are more elaborate and the technique of making them is yet to be fully developed. There are many examples, in the next two chapters, of inferences which may legitimately be drawn from sets of observations of atmospheric pollution. Other inferences, of general and local interest, may be made wherever a sufficient number of systematic measurements have been taken.

CHAPTER 12

Distribution of Pollution

It is important, so long as pollution is allowed to escape into the air, to know the amounts encountered, particularly in streets, houses, and places of work. It is necessary, therefore, to study how atmospheric pollution is distributed. By the methods described in the last chapter, a great many observations have been collected by local authorities and others in Great Britain. At the time of writing of previous editions of this book, few surveys had been undertaken and so conclusions on nationwide pollutant levels were difficult to draw. The classic survey of pollutants in Leicester in 1937–39, published in 1945 and republished in 1956, was used as a basis for suppositions about pollutant levels in other cities. Some information on winter smoke and sulphur dioxide was published in 1950 but co-ordinated and continuous monitoring did not commence until well after the Clean Air Act. The first published report of the Warren Spring Laboratory appeared for the Winter of 1962.

For its historical interest and perspective, the following sections (pp. 147–157) are an abbreviated version of the "state-of-the-survey" before 1962. The final sections of this chapter describe a survey of pollution in Manchester in 1970 and more recently published results from the National Survey.

Historical Perspective
Distribution in Britain as a Whole

Deposited matter

The maps, Figs. 85–88, were prepared from data in the Annual Reports of the Investigation of Atmospheric Pollution and published in the *Quarterly Journal of the Royal Meteorological Society* (1950). They show the distribution of undissolved ash and combustible matter, and of dissolved sulphates and chlorides collected by deposit gauges in country districts well away from large towns or some miles to the west or south-west of the nearest source of pollution. The deposit of ash, undissolved combustible matter, and dissolved sulphates appears to be least in the west and south of England; that of chlorides, which must be due more to sea spray than to chimney pollution, is greatest near the west coast. There was unfortunately no information for Wales, south-west England, Sussex, Kent, East Anglia, or the Scottish Lowlands and Highlands.

The units in Figs. 85–88 are grammes per 100 m² per month. As would be expected,

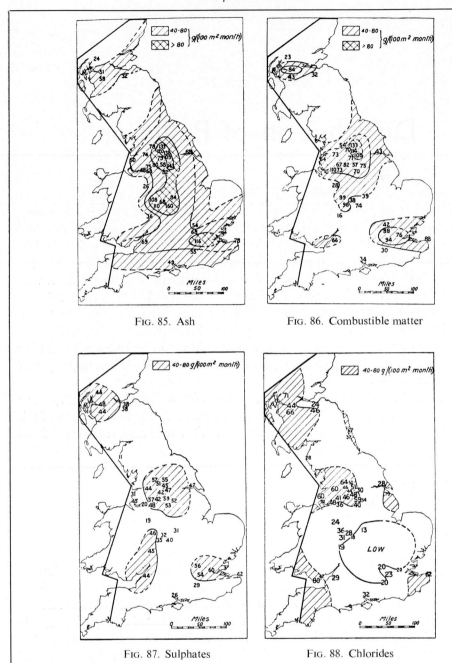

FIG. 85. Ash

FIG. 86. Combustible matter

FIG. 87. Sulphates

FIG. 88. Chlorides

the deposit increases as we go from the country into the suburbs of a town, and increases still more if we enter an industrial area. For example, if the insoluble ash deposits within towns were plotted in Fig. 85, a number of small areas would be shown with deposits over 1000 g/100 m² per month.

The scale of the increase from country to town is not by any means the same for all types of deposited impurity. It is much more rapid for insoluble matter than for sulphates or other forms of dissolved matter. The deposit of ash (i.e. incombustible undissolved matter) in a typical suburb is 15 to 20 times as heavy as in a clean country district, and in the average industrial district it is 2 to 3 times heavier still, while in extremely polluted industrial districts the fall of ash is hundreds of times as heavy as in the country. The urban deposit of dissolved matter is seldom more than double the rural deposit, except where special water-soluble waste products are emitted from industrial chimneys.

These observational results have a simple explanation. Insoluble deposited matter includes a high production of large particles which fall to the ground by their own weight and are seldom carried far from the chimney where they are emitted. Soluble deposited matter is mostly made up of material which passes from the fire into the flue as a gas, vapour, or suspension of very small particles. Thus, much of the soluble matter emitted from urban chimneys is blown into the surrounding country. Some of the soluble impurities in the air, particularly chlorides, come from the sea, since it frequently happens that drops of sea spray are swept into the air during gales. If the air is unsaturated, these evaporate, leaving minute specks of salt which may be blown far inland, where they are equally likely to be caught in a rural or an urban deposit gauge.

TABLE 28. *Pollution deposited in Great Britain*
Average values, in tons per square mile per month
(multiply by 39·2 to convert to g/100 m²/month)

Type of district	Insoluble matter			Dissolved matter		
	Combustible	Ash	Total	Chlorides	Sulphates	Calcium
Country	0·9	1·3	6	0·9	1·3	0·3
Residential suburb	4	5	7	2	2	0·7
Central park	4	6	9	2	3	1
Industrial	7	12	9	2	3	1
Extreme industrial	30	110	30	4	11	10
Pollution emitted (average for G.B.)	3	(2)	—	0·6	10	—

The average quantities of the deposit in various types of district are given in Table 28. The table should be used with great caution in estimating the probable deposit in a single month in a particular neighbourhood, for which purpose it cannot replace direct observation, but it confirms the general conclusions of the preceding paragraphs. The quantities found in extreme industrial districts vary very much, and depend upon the nature of the industry; in particular, the deposit of calcium would not reach the value of 10 tons/sq. mile per month except near lime and cement works.

At the foot of the table are added the rates of emission per square mile per month. They are expressed as averages for the whole of Great Britain east of the thick line in Figs. 85–88, area 65,900 sq. miles. If allowance is made for the area occupied by each type of district, it can be shown that the average weight of ash emitted must be 2 g/(100 m² month) to produce the observed quantities of ash in the deposit. Since this is not an independent estimate, the figure (2) is enclosed in brackets in Table 28. The corresponding total emission of ash is 1·5 million tons/year. The total deposits of combustible matter and sulphur dioxide in Table 28 are much less than the amounts emitted. This suggests either that much of the sulphur dioxide and some of the smoke is blown out to sea or, as has been pointed out in Chapter 11, that the standard deposit gauge may collect less than a fair share of the fine smoke or sulphur dioxide which is deposited in its neighbourhood. An argument that both alternatives occur was put forward by Meetham (1950). The total deposit of chloride is, as might be expected, rather more than the amount emitted from chimneys.

Smoke and sulphur dioxide

Smoke is not easily observed by eye, except from outside and above the smoky area. While the airman's opinion about the distribution of smoke will be approximately correct, therefore, that of the ordinary citizen may be false, being likely to be more nearly the distribution of grit, which is easily observable. It is reported by airmen that smoke is visible sometimes for hundreds of miles downwind from densely populated industrial areas, and many thousands of feet above ground. A photograph taken in 1938 from the American stratosphere balloon, Explorer II, indicated that appreciable concentrations of smoke reach the top of the troposphere whose height varies from 8 to 13 km (26,000 to 43,000 ft) but do not enter the stratosphere. This would be expected, since air mixes vertically within the troposphere, but there is little intermixing with the stratosphere.

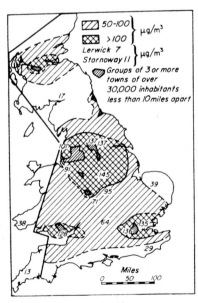

FIG. 89. Winter mean. Smoke in surface country air, mg/(1000 m³)

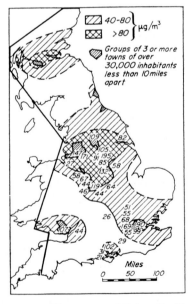

FIG. 90. Winter mean. Sulphur dioxide in surface country air, mg/(1000 m³)

Since 1950 Explorer II's observation has been confirmed by many manned and unmanned flights into the stratosphere.

Practically all measurements of smoke and sulphur dioxide have been made in Britain near the ground. Country air probably contains appreciable quantities at all heights up to several thousand feet, but at moderate heights above a town the air is likely often to be nearly as pure as in the country. Figures 89 and 90 show, as accurately as is at present known, the summer and winter average distribution of smoke and sulphur dioxide in country districts. Although measurements in the towns are excluded, the maps show the effects of industrial areas in polluting the air of rural districts surrounding them.

The above maps were prepared during a study of the life history of pollution over Britain as a whole, including pollution in the upper air. They do not necessarily indicate the "black" areas, but they show where the pollution problem of a town may be aggravated by the influence of neighbouring industrial districts. Perhaps Fig. 89 should be combined in some way with Fig. 91, representing densely populated districts whose locally caused smoke concentration is expected to exceed 100 $\mu g/m^3$.

Figure 91 was taken from *The Investigation of Atmospheric Pollution*, 31*st Report*, 1959, as was the information presented in Table 29. This table gives the range of variation of

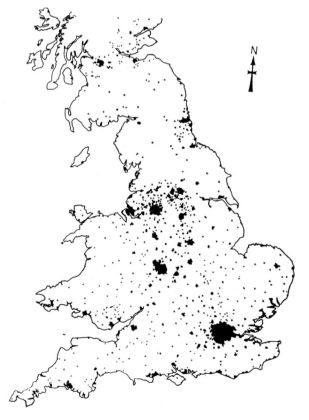

FIG. 91. The average distribution of smoke in Great Britain.
Areas shown in black have an average concentration more than 100 mg/m^3

seasonal average pollution from place to place. The Reports should be consulted for particulars of the places where the measurements were made. The results from 170 or more sites were used for the category described as "average". For the categories "high" and "low" 6–8 sites were used.

TABLE 29. *Smoke and sulphur dioxide in Great Britain (microgrammes per cubic metre)*

	Average smoke		Average sulphur dioxide	
	Summer	Winter	Summer	Winter
High	380	710	257	485
Average	120	270	143	257
Low	10	20	9	57

The densities of smoke and sulphur dioxide are about the same at similar sites, but the variation from High to Low is different: nearly 40–1 for smoke, under 10–1 for sulphur dioxide. The density of pollution at ground level depends on other factors beside the rate at which it is emitted from chimneys. Two such factors are not necessarily the same for smoke as for sulphur dioxide: (1) the height of the chimney from which the pollution emerges, and (2) the rate at which the pollution is deposited or otherwise removed from the air. It is evident from Table 22, Chapter 10, that an undue proportion of smoke comes from domestic chimneys, which are much lower as a rule than other chimneys, and it also seems probable that smoke is removed from the air less quickly than sulphur dioxide. Both these facts help to explain why the density of smoke near street level is more than that of sulphur dioxide. (The present tense of these verbs is only the "historical present.")

Most British households have open grates in which coal is burnt, and the quantity of smoke produced in a town is roughly proportional to its population. But because larger populations usually occupy larger areas of land, the average density of smoke in the centre of a town does not increase in direct proportion to the population. In non-foggy conditions, local density of housing is a more important factor than total population of a town. This has been brought out clearly by J. Pemberton *et al.* in Sheffield: in this hilly, heterogeneous city they found a clear linear relation between the number of electors per acre in various electoral wards (of very different densities of population) and the smoke concentration in each ward. The same was not true for sulphur dioxide; its concentration must depend much more on the fuel used by the main industries within several miles of the point of observation.

Smoke and sulphur dioxide travel long distances across country, as can be inferred from Figs. 89–90. In rural districts near Leicester the quantities of suspended impurity were found to depend strongly on the direction of the wind. When a large group of observations was analysed, it was shown that the quantity of pollution depended on the size and distance upwind of the nearest densely populated region. For distances between 35 miles (the Birmingham district) and 90 miles (the London area) the pollution in

Leicestershire was proportional to the population of the industrial area and to the inverse square of the distance. The inverse square law is to be expected if, during its journey, the pollution diffuses upwards and sideways.

Distribution within a Town

Deposited matter

Although Table 28 gives some idea of how much deposited pollution to expect in different quarters of a town, it gives no clue to the detailed distribution. In most towns this must be extremely intricate, since round many factory chimneys there will be regions where deposits are heavier than normal. Pioneer work on the distribution of insoluble matter was conducted in Bilston, Staffordshire, by the Borough Health Department. From observations by the petri-dish method it was shown that there were real differences in deposited matter at places only a quarter of a mile apart; and hence that, in certain weather conditions at any rate, solid matter is deposited mainly within a a quarter mile of its place of origin. The map reproduced as Fig. 74, p. 135 showed that deposits in some districts of Bilston were more than 10 times as heavy as in others only a mile away.

Smoke and sulphur dioxide

Figures 92 and 93, representing the yearly average distribution of smoke and sulphur dioxide in London in 1957–8 are taken from a paper by T. S. Pindard and E. T. Wilkins (1958). A key to the names of the boroughs and the positions of open spaces and smokeless zones is given in Fig. 94. It is noteworthy that the smokeless zone marked "City" appears to have caused a well-marked reduction in the smoke concentration in the City of London area.

In the published report of the Leicester survey, there is much information about the general distribution of smoke and sulphur dioxide in a town, and about the effect of wind on the distribution. Figures 95 and 96 show the average distribution of smoke and sulphur dioxide in Leicester, in winter, and this may be regarded as typical of what is occurring in other towns. The closely built-up area is represented by hatching, and the areas covered mainly by houses with gardens are stippled. Because of the relatively few points of observation, contours of equal pollution have been drawn as ellipses; even so, they conform fairly closely to the shape of the built-up area.

Apart from a relatively small asymmetry due to the prevailing westerly winds, the average distribution of pollution in Leicester was little different from the distribution of the chimneys emitting it. The factories and central-heating installations near the middle of the city apparently produced considerably more sulphur dioxide than smoke, and the distribution of sulphur dioxide in the figure shows a higher maximum at the centre than the distribution of smoke. The ratio of the concentration in the centre to that in the built-up suburbs was 3·4 to 1 for sulphur dioxide and only 2·2 to 1 for smoke.

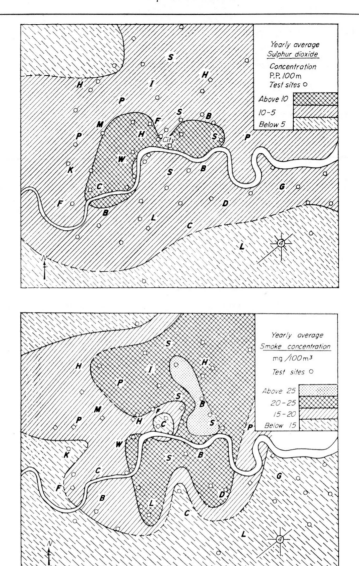

FIG. 92 AND 93. Yearly average distribution of smoke and sulphur dioxide in London in 1957-8

The effect of wind on the distribution of pollution was analysed by grouping the observation in Leicester according to the speed and direction of the wind. It was also possible to estimate the average pollution in the air entering the area, and so to determine approximately the amount of native Leicester pollution.

The observed result, that the centre of Leicester nearly always had the highest density of pollution, is in agreement with Fig. 95, which indicated a reduced yearly average concentration of smoke in a smoke-controlled zone only about 1 sq. mile in area. The results suggest that suspended or gaseous pollution must be rapidly removed from near street level; but removal sideways is out of the question near the centre of a large town, and

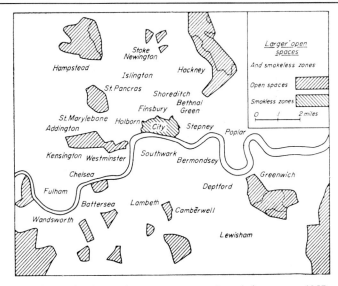

Fig. 94. London boroughs, open spaces and smokeless zones, 1957.
After T. S. Pindard and E. T. Wilkins, Conference of the Nat. Soc. for Clean Air, 1958

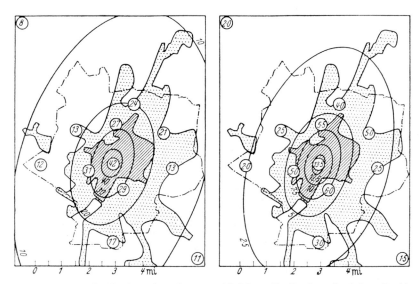

Fig. 95. Mean distribution of smoke. Concentration in mg/100 m^3

96. Mean distribution of sulphur dioxide. Concentration in parts per 100 million

removal downwards—i.e. deposition on the ground—takes place too slowly to produce the observed result. So it seems that the pollution escapes upwards. Until samples of pollution are taken at considerable heights, this conclusion can best be tested by measurements of daylight at the ground, since smoke above the chimney tops cut off just as much daylight as smoke nearer the ground. In the right-hand half of Fig. 97, the winter-average losses of daylight due to smoke are given. The maps show quite clearly that the greatest loss of daylight occurs at a point downwind from the centre, usually about a mile. This confirms the conclusion that smoke rapidly escapes upwards from street level. Beyond a mile downwind, it seems likely that the pollution spreads sideways sufficiently for the total weight of smoke over unit area of ground to begin decreasing.

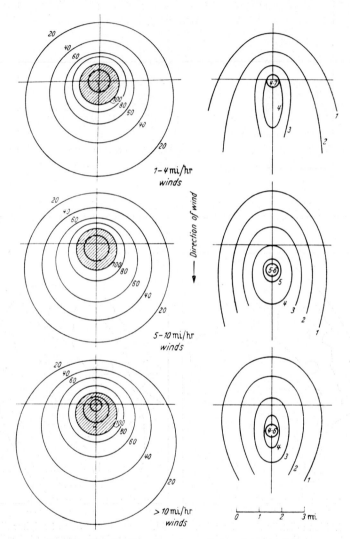

FIG. 97. Distribution of smoke and ultraviolet daylight

From the practical point of view, the upward diffusion of smoke and sulphur dioxide is particularly important, because it is the main factor in preventing the accumulation of pollution in disastrous concentrations. If we were foolish enough to build a roof 100 ft high over one of our cities, shortening any tall chimneys to less than 100 ft, the concentration of sulphur dioxide would be 15 mg/m³ about 1 hr after the fires and furnaces were lit, and the density of smoke at the same time would be about 5 mg/m³. Even if there was a wind blowing, these amounts would be reached in leeward districts whenever the wind took more than an hour to pass right through the town. And if, as often happens, the wind dropped below 1 or 2 miles/hr, the concentration of sulphur dioxide would exceed 30 mg/m^3, which some authorities regard as dangerous to even healthy persons. That such concentrations rarely occur in towns is due to the diffusion of polluted air upwards and of clean air downwards. More is said about turbulence and diffusion in the next chapter.

Recent Surveys

Both data collection and interpretation for the urban surveys which were conducted either before or immediately following the first Clean Air Act (1956) were simplified by two important factors. Firstly, the levels of both smoke and sulphur dioxide were so high that small sampling and observational errors were negligible and, secondly, the surveys (especially those in Leicester by Meetham and in London by Chandler) were, perhaps forlunglely, undertaken for areas with comparatively simple topography. More recent studies have included cities for which the air pollutant loading is strongly modified by the local topography—for instance, the "bowl-shaped" urban area of Leeds and the complex terrain in and around Sheffield (see Garnett, 1967).

FIG. 98. View of Manchester. (Courtesy of Studio Life.)

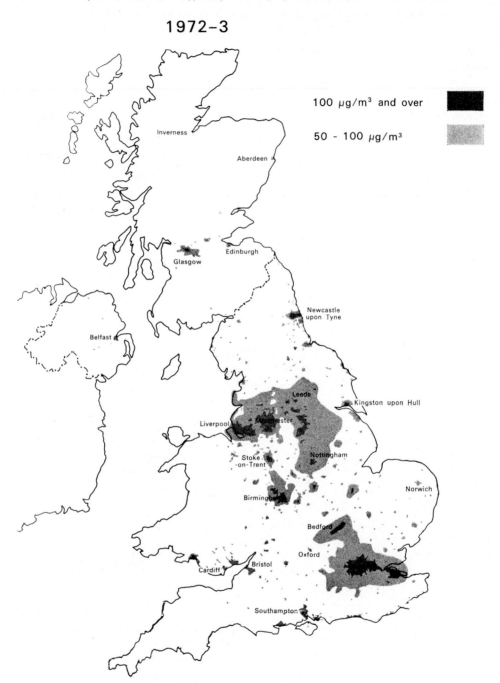

FIG. 100. (Reproduced with permission from a map produced by the Department of the Environment)

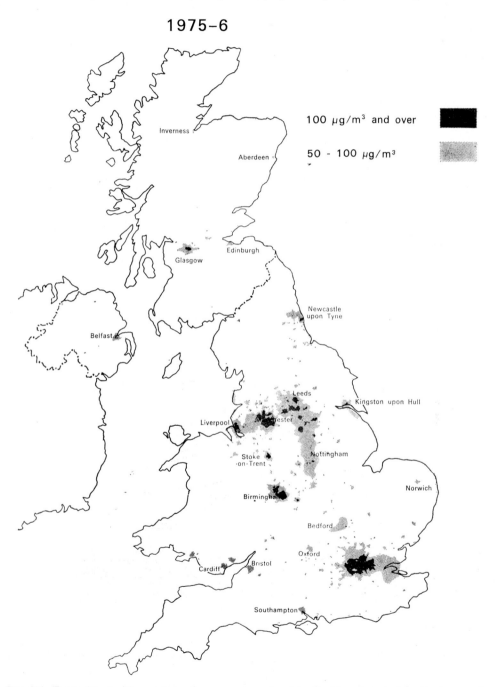

FIG. 101. (Reproduced with permission from a map produced by the Department of the Environment)

Atmospheric Pollution

The pollution in many of the northern cities of the U.K. was extremely bad in the 1940s and 1950s. This perhaps renders the contrasts between "then and now" all the more pleasing! The County of Greater Manchester is one such success story as the publication *20 Years of Air Pollution Control* illustrates. Levels have dropped across the county from around 400–500 μg m^{-3} in the late 1950s to present-day values of less than 100 μg m^{-3}. For the citizens of this area, however, the view (see Fig. 98) and the cleanliness of washing, windows and indeed white chrysanthemums are the real and worthwhile benefits!

The National Survey of Air Pollution

Since the Clean Air Act of 1956 and the inception of the Warren Spring Laboratory, there has been a concerted (and growing) effort to monitor air pollution and evaluate spatial differences and yearly trends. Members of the Standing Conference of Cooperating Bodies report smoke and sulphur dioxide values (24-hour means) on a continuing basis. Some members are asked periodically to participate in smaller-scale experiments (e.g. survey of sulphates, April 1976–March 1978). (Other surveys are carried out—for instance, the AERE Harwell survey of various heavy metals.)

Smoke

Smoke emissions have been estimated for domestic and industrial sources. Figure 99 shows the decreasing trend in both components since 1950—a direct result of the Clean

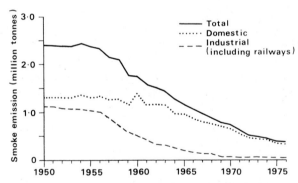

FIG. 99. Trends in smoke emissions (U.K.). Source of data: Warren Spring Laboratory Reports

Air Act. The trends in resultant observed concentrations, taken mainly from urban sites, are shown in Fig. 105 (Chapter 13). Geographical variations are shown in Fig. 100 and 101 highlighting the continuing high concentrations in our major conurbations.

Sulphur dioxide

Trends in sulphur dioxide emissions over the same period (see Fig. 102) show that domestic and industrial emissions have fallen and remained approximately steady respectively; but power-station emissions are still increasing. Concentrations of sulphur dioxide (from the Warren Spring reports) are seen to be falling in urban areas (see Fig. 105 in Chapter 13) although the mean U.K. value for 1976 of 62 μg m^{-3} is still just above the WHO recommended annual mean value of 60 μg m^{-3} (cf. recommended level for smoke of 40 μg m^{-3}). However, in many large urban areas (notably London, Yorkshire, the North-west and the Midlands) mean values are still well in excess of these guidelines.

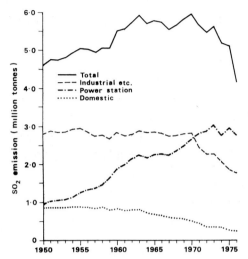

FIG. 102. Trends in sulphur dioxide emissions (U.K.).
Source of data: Warren Spring Laboratory reports

Concluding remarks

With the continuing trend of decreasing levels of smoke and sulphur dioxide, studies of urban areas have widened both in terms of the number of pollutants considered and the approach adopted. It is now recognized that Man's pollution of his built environment includes the considerable temperature modifications—generally called the "urban heat island effect"—which is caused by the combination of a number of factors including modified albedo and run-off characteristics as well as the obvious dispersal of "waste" heat to the atmosphere. However, more recently, it has been shown that the local meteorology is also considerably perturbed by the very presence of a conurbation. The perturbations caused in cloud formation, atmospheric absorption, wind velocity and turbulence and precipitation feed back directly into the urban environment through the alterations in dispersal patterns of pollutants.

The large local meteorological modifications described above and the very much lower levels of smoke and sulphur dioxide combine to make analysis and prediction of pollutant levels an increasingly complex task. Air-quality-control programmes are designed and implemented on a county-wide basis but it has been suggested that the, somewhat artificial, nature of the boundaries may lead to difficulties (Henderson-Sellers, 1979) especially in view of the increasingly global nature of air-pollution problems (OECD report on Long-range Transport of Air Pollutants, 1977).

CHAPTER 13

Variability of Pollution

THE study of atmospheric pollution is admittedly an untidy science. It is analogous to some aspects of history, economics, medicine and meteorology, because it cannot be properly developed by the neat experimental methods which are so successful in physics and chemistry. Much of what has been learned about atmospheric pollution has been acquired by making series of observations and examining them after a large number have been collected. If a set of observations shows some unusual trait, a tentative explanation is offered, and this is tested by making other sets of observations, often quite different in character from the original set.

As an example of such a process of thinking, it has been suggested that smoke and sulphur dioxide must be removed from the air by natural processes; and to be consistent with observations of the atmospheric pollution in Britain it has been estimated (Meetham, 1950, 1954) that the average life of a smoke particle before deposition on land must be 1 to 2 days, and that of a molecule of sulphur dioxide less than 12 hr. Revised estimates for the residence time of sulphur dioxide are of the order of 4 days.

Further work is needed for the above assertions to be checked or replaced. There is indeed seldom much difficulty in offering an explanation; the difficulty is normally to decide which of a number of explanations is nearest the truth, and to obtain confirmation by demonstrating that the unusual trait is without exception the consequence of the particular cause offered in the explanation.

The techniques of statistics are an invaluable help in studying sets of observations of atmospheric pollution. They can be used for testing the closeness of the relation between suggested cause and effect, by means of the correlation coefficient; by this means some of the conclusions near the end of the present chapter have been reached, about the effect on atmospheric pollution of wind, temperature, rain and other factors.

In addition, statistics can be used to prevent undue waste of time in investigating changes in pollution which are due to a chance concurrence of many different causes, and also to ensure that the maximum information is extracted from each set of observations. Anyone who makes regular observations of atmospheric pollution will have experienced the difficulty of "knowing where to begin" in analysing the lists of results—and, still more the difficulty of knowing when to stop. The particular branch of statistics which is used for these purposes of control is known as "significance testing". Briefly, the method is to estimate how large a change is likely to occur as a result of chance concurrences of different causes, and to pay further attention only to those changes which are too large to be of this kind, i.e. to those changes which are "significant". For example, in Fig. 103 it might be thought that some real change in the emission of pollution must have caused the succession of 9 months when the total insoluble matter was equal to or below the average, represented by the smooth curve; but the degree of scatter of the individual points in the figure is so great that it can be shown, by significance tests, that the succession could easily be due to chance concurrences of the same characteristics (perhaps variations of weather) which produced the scatter.

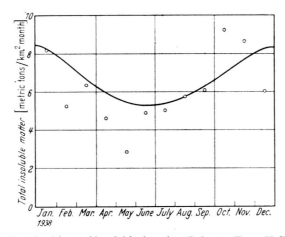

Fig. 103. Monthly total insoluble deposit at Leicester Town Hall (1938)

By a study of the significant changes in atmospheric pollution, it has been possible to attain a clearer understanding of both human and other factors which affect pollution. A general example may help to illustrate the method.

Suppose regular observations at a fixed point indicate that there has been a change in pollution. We infer that there has been a change in either (1) the rate of emission of pollution from one or more of the sources from which it can reach the point of observation, or (2) the mechanism by which the pollution from these sources is brought there. The mechanism is mainly a matter of weather—wind speed and direction, atmospheric turbulence, rain and so on. We can eliminate the effect of any change in the effect of weather by averaging observations over such a long term that all types of weather are represented in about their normal frequency. The longer the term, the more nearly is weather eliminated; and if the change in pollution is still evident, the more certain is it, according to tests of significance, that there has been a change in the average emission from the chimneys whose pollution can reach the point of observation. It is usually more important, however, not to wait until we can be absolutely certain, but to be reasonably certain in the shortest possible time. In this case statistical tests of significance are indispensable, since they permit degrees of certainty to be expressed in terms of mathematical probability. A probability of 95 per cent or more is usually regarded as "significant".

To continue the example, suppose the change in pollution was due to a change, not in the amount emitted, but in the weather. This could be part of a periodic change, such as the seasonal variation of weather from winter to summer, or it could be an irregular happening, for example the persistence of easterly winds for two exceptionally long spells during 1947. We investigate periodic changes in the same way as we investigate changes in emission, by averaging a large number of observations. For instance, from the averages of the observations in a large number of Januarys, Februarys, etc., separately for each month, we can deduce the average seasonal variation of pollution. To investigate the effect of a change of wind or any other irregular change, we collect observations until the change has been repeated many times and then determine the coefficients of correlation and regression. These are the general methods, in outline, of examining a change in the pollution at any point of observation. Particular examples will now be considered of changes which have been shown to be significant.

Changes in Deposited Matter

Of all the instruments for measuring pollution, the deposit gauge has been longest in regular use. At a number of places in England and Scotland, regular monthly observations have been made since 1914. Though these give but little information—and that only indirectly—about the complicated day-to-day fluctuations in the rate at which pollution is deposited, they are useful for studying the trends and the yearly cycle of deposited matter.

The trend of pollution may be defined as its tendency to increase or decrease steadily as time goes on. Observations are usually needed for several years to verify that a steady change in pollution is taking place but, once conclusively established, a trend in atmospheric pollution implies that there is a similar trend in the rate at which pollution is being emitted. It is thus possible to tell from observations whether in any neighbourhood the emission of pollution is increasing or whether, on the contrary, effective steps are being taken to reduce it.

TABLE 30. *Trends of deposited matter:—Example Insoluble ash*

Period April–March	London Finsbury Pk.	Rothamsted (Herts)	Sheffield Attercliffe
	average as tonnes per square kilometre per month		
1914–1919	5·9	—	—
1919–1924	3·4	1·0	—
1924–1929	3·6	1·1	—
1929–1934	2·9	0·9	4·3
1934–1939	2·5	0·7	5·4
1939–1944	1·7	0·5	7·3
1944–1949	1·9	0·5	7·5
1949–1954	2·6	0·5	6·4
1954–1959	2·1	0·4	5·4
	Decreasing to 1944	Decreasing	Increasing to 1949

Examples are given in Table 30 of 5-yearly averages of deposited matter for each of which up to sixty observations were made. Tests have shown that, in the particular examples quoted, the minimum change which could be accepted as significant is one of between 20 and 50 per cent; thus all the trends indicated in the table are significant. It must however be remembered that the deposit at any one situation is not necessarily typical of a whole town. Instances have occurred of the deposited pollution increasing at one situation while decreasing at another within a few kilometres. The pollution collected by a deposit gauge, particularly the undissolved solid matter, is chiefly of local origin.

Yearly cycle

The average rate at which pollution is deposited is in many places greater in winter than in summer (Fig. 104), although the differences are seldom large, and there are districts where the reverse holds true. After five or more years' observations, if the average rate of deposit in summer is divided by the average winter rate, the ratio is almost certain to be between 0·5 and 2·0.

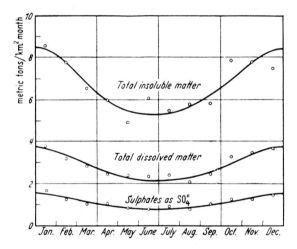

FIG. 104. Yearly cycle of deposited pollution at Leicester Town Hall, 6 years ended March 1939

Changes in Smoke and Sulphur Dioxide

Whenever a local authority establishes one or more smoke-control areas, the question is asked "What good has it done; how much smoke have we got rid of?" The answer cannot be given quickly. Many months or years of measurement are needed both before and after the change. Allowance must also be made for peculiarities of weather.

Craxford (1961) gave tables and diagrams showing the changes in smoke and sulphur dioxide at all British observing stations between 1952 and 1960 (see Fig. 105). In this period the number of stations increased from 47 to 266 for smoke, and from 28 to 219 for sulphur dioxide. The Clean Air Act became law in 1956, and there was a rapid expansion in the installation of smokeless domestic appliances, and central heating systems, with a corresponding reduction in industrial and railway smoke.

During this period, only the initial effects of the Clean Air Act can be seen. Over 20 years after the 1956 Act, the total reduction in smoke can only be described as "dramatic" (less so for sulphur dioxide). Figure 105 reveals a decreasing trend in both these pollutants. Mean smoke concentrations are now below the WHO recommended standards although further reduction (especially of sulphur dioxide) is both desirable and necessary.

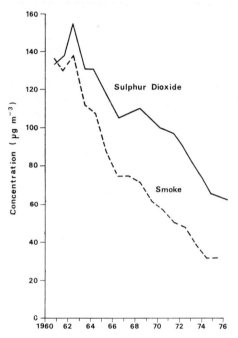

FIG. 105. Trends in average urban concentrations of smoke and SO_2 in the U.K.
Source of data: Warren Spring Laboratory reports

The Yearly Cycle of suspended pollution is much more definite than that of deposited matter. At most places the average concentration of smoke and sulphur dioxide in winter is 2–3 times the summer value. This is partly because of the extra fuel used in winter in nearly every locality, and partly because atmospheric conditions are more favourable in summer for the dispersal of smoke and gases away from street level into the upper air. Examples of yearly cycles are given in Fig. 106.

The Weekly Cycle of smoke and sulphur dioxide is characterized in urban districts by lower concentrations during the week-end, when the concentration may fall by from 20 to 40 per cent. Few observations from residential districts are available, but here a week-end rise might be possible. Regular variations between Monday and Friday are usually slight.

Because of the yearly cycle, it is better to study the weekly cycle of summer and winter observations separately; and because of the vagaries of weather, the weekly cycle can be

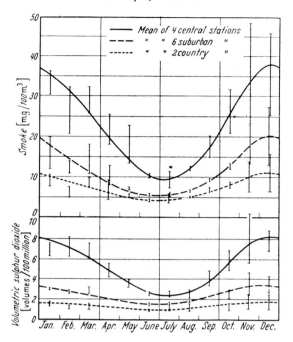

Fig. 106. Yearly cycle of suspended pollution at Leicester, 1937–9

adequately studied only from large groups of observations, covering several years. Examples are given in Figs. 107 and 108. Although over a long time period there is a detectable difference in the meteorology of conurbations between week-ends and weekdays due to variations in anthropogenic heat output (and the resultant changes in the structure of the heat dome), the lower concentration of pollution at week-ends in Figs.

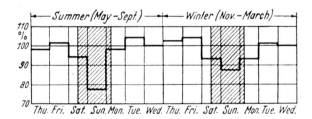

Fig. 107. Mean weekly cycle of smoke, Stoke-on-Trent. Leek Road, 1939–44

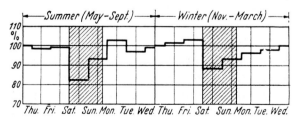

Fig. 108. Mean weekly cycle of sulphur dioxide, Sheffield, Surrey Street, 1939–44

107 and 108 is still thought to be largely attributable to a diminution of pollution at week-ends.

The Daily Cycle of suspended pollution is one of its most interesting features because, as a rule, there are two maxima each day. When domestic coal burning was prevalent

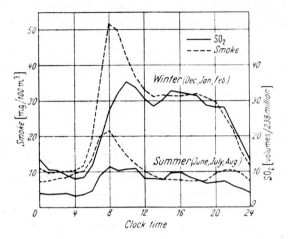

Fig. 109. Daily cycle of smoke and sulphur dioxide

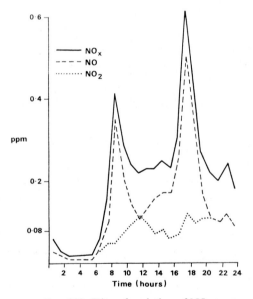

Fig. 110. Diurnal variation of NO_x

this could be clearly seen as a maximum at 8 a.m. and again one at 6 p.m (see Fig. 109)— the corresponding maxima for sulphur dioxide occurred at 10 a.m. and 3 p.m. The difference between the intensities of the morning maxima was due to a fundamental difference between smoke and sulphur dioxide: sulphur dioxide is produced whenever coal is burnt; whereas smoke is produced only when combustion conditions are bad, and they were worst of all when fires were being kindled in cold grates. The examples given indicate that in the 1940s and 1950s about half the smoke at 8 a.m. in Leicester was produced by the kindling of fires in cold grates.

In the 1970s these twin peaks are now seen in another pollutant: NO_x. Emissions correspond to traffic peaks at approximately 9 a.m. and 6 p.m. (see Fig. 110), mimicking the smoke and sulphur dioxide curves of Fig. 109 and reflecting our change of habits and life style.

Irregular Variation

Irregular changes in almost any pollutant can occur causing concentrations of possibly more than 20 times the average (measured on a daily basis) and variations at a given time may be several orders of magnitude higher than the daily average, although such conditions seldom persist nowadays for long periods (cf. the London smogs of the 1950s).

Because suspended pollution can conveniently be measured over daily or hourly intervals, its irregular variation has been studied more closely than that of deposited matter. At any point which is not directly downwind of individual chimneys, the suspended pollution usually has come from an area of several square kilometres or even more. Casual changes in the rate of firing individual furnaces and fires in such a large area might cause slight differences in the observed concentration of pollution, but they could not conceivably cause differences so great as 10–1. Even systematic changes in the rate of firing, such as might be caused by a change of temperature, could not produce such differences in pollution. The chief causes of irregular variation in the density of suspended pollution are in fact meteorological. They will now be briefly considered in the order: direction of wind, temperature, rainfall, velocity of wind, and turbulence.

Air Pollution Meteorology

Direction of wind is the only meteorological variable which is likely to affect differently the pollution in different districts. Changes in wind direction cause the greatest fluctuations in pollution in suburban districts, because of the increase produced when the wind begins to blow from the centre of the town, having previously blown from some other direction.

The amount of fuel burnt for space heating is of course dependent on the *temperature* out of doors, and the yearly cycle of pollution is in part a consequence of the yearly cycle of temperature. Irregular changes in temperature are a common feature of the British climate, and it is not surprising that there are corresponding irregular changes in atmospheric pollution, the denser pollution being associated with the colder weather. This proves to be the case, and in the winter a change from the warmest to the coldest weather may more than double the concentration of sulphur dioxide (daily mean value). Over a period of 24 hours it is not uncommon for urban concentrations of sulphur dioxide to vary by a factor of 10 or more.

Rain plays an important part in cleaning the air of smoke and sulphur dioxide, though from existing observations it is difficult to form an estimate of how much of the irregular variation of suspended pollution is due to the spasmodic way in which rain falls. It is probably fair to say that rain in summer or winter has about as much influence as temperature in winter; on a warm dry day in winter the suspended pollution is likely to be equal to that on a cold wet day in summer.

The velocity of the wind might be thought to be by far the most important factor in causing irregular variations in every kind of atmospheric pollution. There is no doubt that it does control the distance travelled by large particles before they reach the ground, but in its action on suspended pollution near ground level, wind velocity is less important than might be supposed. Provided that the speed with which smoke and sulphur dioxide spread upwards is unchanged, alterations in wind speed do not strongly affect their density near ground level. In a large town, for instance, a mere increase of wind means that pollution is kept close to the ground for a longer distance from the chimney, and the fact that more air is passing the chimney tops is offset by the decrease in height to which pollution spreads above the town. In the centre of Leicester, the density of pollution in barely perceptible winds of 0.25 m s^{-1} was found to be from 2 to 4 times the density in winds of 10 m s^{-1}, provided the turbulence in both cases was normal.

It is seldom the case that the wind blows consistently in one direction. Lateral wind velocities (i.e. components) exist which cause sideways, upwards, and downwards motions of pollutants placed in the wind field. These motions are usually of a random nature, associated with the eddy motions in the air. This *turbulence*, if present, ensures that pollution is diluted rapidly. In an urban area, the various sources contribute to a fairly well-mixed air mass. Upward spreading will slowly increase the depth of this layer of pollution; downward spreading will eventually result in the pollution reaching ground level. *Ground level concentrations* (g.l.c.) are perhaps of greatest concern.

Most of the time there is enough eddy motion in the air for suspended pollution to diffuse upwards from towns and be partly replaced by cleaner air from above. There are times, however, when turbulence is nearly absent, and while there may be sufficient turbulence for pollution to diffuse down to ground level from the chimney tops, the upward diffusion may be very slow indeed. As a result of changes in turbulence which commonly occur, the density of pollutants near the ground may change by as much as 10-1.

There tends to be little turbulence during calms or when the wind is light, because few eddies are then produced by trees and buildings in the path of the wind. Turbulence also tends to be less at night, because the ground is then often cooler than the air above it, and there is no possibility of turbulence due to local heating of the air. On cloudless nights, when the ground cools most rapidly, the surface layers of air cool also, and a condition is set up in which the air is warmer above than near the ground. This condition is known as an *inversion* (more strictly, a surface inversion). Later, the region of the inversion may rise from the ground, and the temperature of the air may vary with height as in Fig. 111. Where there is an inversion, vertical eddy motion due to local heating of the air is constrained, and in addition, vertical eddies of any kind die out more quickly than usual.

There is least turbulence of all, therefore, during calms or light winds accompanied by an inversion. These conditions occur most frequently at night, especially in autumn and winter, and they tend to persist into the morning. This is why, when furnaces are fired in the early morning, they often produce the greatest levels of pollution. High concentrations may also occur in these weather circumstances in the evening after sunset.

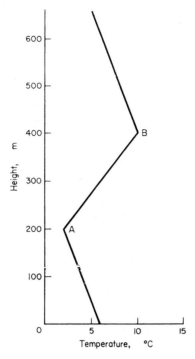

FIG. 111. Variation of temperature with height in a fog

Chimney plumes

The effects of meteorological variations on pollution can often be observed directly by watching the emission of pollutants from a chimney. The plume of material issuing forth from the chimney in steady neutral conditions (i.e. an environmental lapse rate equal to the adiabatic lapse rate) forms a cone, usually visible nowadays because of its water content (in the past, plume visibility was more strongly determined by the particulate loading, i.e. smoke). An instantaneous plume tends to have ragged edges, but a time-exposure photograph shows a smooth shape. Measurements of variables such as pollution level, buoyancy (i.e. excess heat content) and velocity across the plume reveal a Gaussian (or normal) distribution (see Fig. 112). Mathematically the so-called "Gaussian plume model" is useful to predict downwind concentrations, both in the atmosphere and when the plume widens (and dilutes) sufficiently to touch ground level (see later discussion on ground-level concentrations). The rate of spread depends upon the degree of turbulence and, to some extent initially, on the temperature and efflux velocity of the effluent.

Emissions of low velocity may be "sucked down" into the low-pressure region in the lee of the stack *(downwash)* or be trapped in the boundary layer or wake associated with a high building *(downdraught)*. These phenomena necessitate the use of tall stacks to minimize ground-level concentrations.

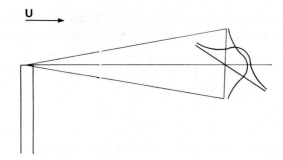

FIG. 112. Gaussian distribution holds across plume cross-section in any direction

A plume with sufficient buoyancy will, however, rise well above the stack top before the wind effectively bends it over. The extra height *(plume rise)* is often expressed in terms of an *equivalent chimney height*—the height of a chimney emitting a plume with no buoyancy or vertical momentum such that the ground-level concentrations are identical. Enhancing plume rise or predicting the height to which a given effluent will ascend is thus of direct benefit in minimization of air-pollution problems; although it must be noted that the plume is also strongly influenced by the vertical temperature structure of the atmosphere and the level at which pollutants are emitted with respect to inversion layers.

Fogs

A natural fog consists of droplets of water suspended as an aerosol. Any particulate pollution in the atmosphere provides extra condensation nuclei and both particulate and gaseous pollutants can be trapped in the resultant stable air mass. High pollutant concentrations emitted in a natural fog can form a "smog"— a term coined by Dr. Des Voeux in 1905. Many of the major pollutant episodes, responsible for increased mortality, have been associated with fogs—indeed the smog in London of December 1952 (see later discussion) led to the formation of the Beaver Committee whose recommendations were embodied in the Clean Air Act of 1956. Although nowadays fogs are seldom converted into smogs, it is useful here to consider the physics of fogs—if only in deference to the important role smog played in prompting Clean Air legislation.

When the surface air is cooled by contact with the ground, its temperature may be reduced below the dew point, and dew or frost settles on the ground. If, however, a thick layer of air is cooled below the dew point, water condenses within the air to form extremely small droplets, and a fog is produced. Fogs occur most frequently in the early morning, during calms or light winds accompanied by an inversion, and it should be noted that these are the very conditions which produce high concentrations of pollution.

A well-established fog may cover a wide area and extend to several hundred metres above the ground. Sunshine falling on top of the fog is reflected away, so that little heat reaches below to help in the evaporation of the water droplets. Below the fog top, at which there is always an inversion, the temperature of the air often changes normally with height as in Fig. 111 below level A. This temperature gradient indicates that mixing is taking place, and indeed fog may be often seen to be swirling, usually very slowly. Unfortunately, mixing of the fog and air has the effect that gases and smoke from even tall chimneys can diffuse down to ground level, although they are prevented from escape upwards by the inversion at the top of the fog.

Constituents of the London fog, December 1952

When the smog catastrophe occurred on 5–9 December 1952 routine daily measurements of smoke and sulphur dioxide were being made at 12 sites in the London area, and monthly measurements of sulphur dioxide at 117 sites. This was the first occasion, apart from November 1948 in London, when any measurements of respirable pollution were made during a fatal smog. Figure 113 has been redrawn from a figure in E. T. Wilkins'

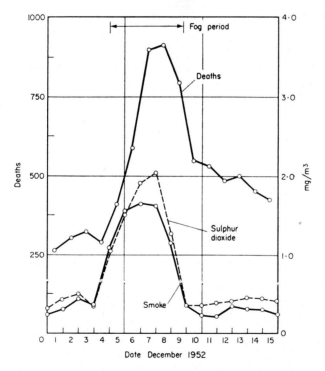

FIG. 113. Daily air pollution and deaths, after E. T. Wilkins, "Air pollution and the London fog of December 1952", *J. Roy. San. Inst.*, 74, 1–21, 1954

account of the disaster given to the Royal Sanitary Institute (now known as The Royal Society for the Promotion of Health). These, however, leave us ignorant of the detailed distribution and composition of the fog, and particularly of its sulphuric acid content. Fog particles have subsequently been studied with the electron microscope, and found to be extremely complex. Examples of high pollution, including sulphuric acid, have also been found when there was no fog.

A surprising feature of the measurements in 1952 was that, although the concentrations of pollution were the highest ever observed in the open, the daily averages were nowhere more than 10 or 12 times the typical December mean. Higher concentrations than these averages must have occurred in patches, and some pedestrians reported having passed through regions which made them cough violently.

There remains the possibility that some other form of pollution, or some combined (or *synergistic*) effect of different forms of pollution, undetectable by the standard measuring instruments, may have caused the fatalities.

The only other possible culprits, about which a little evidence (mainly inferential) is available, seem to be sulphur trioxide, sulphuric acid, carbon monoxide, hydrochloric acid and fluorides, and carbon dioxide as an accessory. All these materials are formed during the combustion of coal, coke, or smokeless solid fuel and have been described in detail in Chapter 10.

There were no direct observational data about the concentrations in the London fog of any forms of pollution except smoke and sulphur dioxide, but it is possible to estimate the daily amounts emitted in the London area. It is also possible, within an error of -50 to $+100$ per cent, to estimate the volume or mass of air with which they were mixed. If, in addition, plausible assumptions can be made about the time they remained in the air, an estimate of their concentrations is possible. No such estimate would be acceptable unless it were consistent with other factors, such as the heat balance and water balance of the fog, and the following paragraphs (quoted verbatim from Dr. Meetham's account of the smog published in earlier editions of this work—units left in imperial) are an attempt to present a consistent account of the London fog of December 1952. Some of the figures have since been revised but they are interesting as they stand here and thus deserve consideration.

Quantities

Consider a rectangular area including Edgware and Chingford in the north, Hornchurch and Dartford in the east, Croydon and Surbiton in the south, Hounslow and Pinner in the west. This area of 450 sq. miles lies mainly in the bottom of the London Basin and must have a population of about 8 million. The fog covered the whole of the area its and height was observed to be up to 500 ft. The mass of air in this volume was 226 million tons; its temperature was little over 0°C.

The meteorological conditions were such that the air could not escape upwards beyond the height of 500 ft; nor could the impurities diffuse above this level at any effective rate. The horizontal winds were so slight and variable in direction that there was probably less than one complete air change in the region in 4 days. It is, therefore, possible to consider the fog region as a closed system, like an enormous shed with all its doors and windows closed.

The air contained about 2,000,000 tons of liquid water (fog droplets) and 750,000 tons of water vapour. At a very rough estimate, the film of water on the ground, vegetation and other objects, weighed 500,000 tons (D). [N.B. The letter (D) is used to indicate any estimate which may be even less accurate than the error of -50 to $+100$ per cent suggests.] The air also contained 380 tons of free smoke and 370 tons of free sulphur dioxide, not including any that was attached to or dissolved in fog droplets, or any sulphur dioxide attached to smoke. No other impurities were directly measured.

Each day about 70,000 tons of coal were burnt in the region; 1000 tons of smoke particles, 2000 tons of sulphur dioxide, 140 tons of hydrochloric acid, and about 14 tons of fluorine compounds were emitted. From the imperfect combustion of coal, and from motor vehicles, 8000 tons/day of carbon monoxide were emitted. The total emission of carbon dioxide was 200,000 tons/day, and the air originally contained 90,000 tons of carbon dioxide.

Heat balance

Practically all the heat from the burning of 70,000 tons of coal per day went into the fog region, and it was surrounded on all sides by air and earth that was warmer than itself. Indeed an amount of heat equivalent to the burning of another 14,000 tons of coal was daily conducted into the fog from the ground. The bodies of the inhabitants contributed the heat value of about 1400 tons of coal, and there was an unknown contribution from the sun, which shone brightly each day on the top of the fog, but most of its heat was reflected rather than absorbed. Together, this incoming heat was enough to warm the whole fog region by over 10°C per day, and yet it hardly altered at all in temperature during the 5 days of its existence. How was such a region, surrounded as it was by warmer objects, able to lose so much heat that it remained cold? The answer, according to geophysicist, is concerned with the power of water and water vapour to emit infrared radiation, while not absorbing radiation of other wavelengths. Indeed, the behaviour of a fog region in sunlight is exactly opposite that of a glasshouse. Calculations can be made showing that the loss of heat was the right order of magnitude to keep the fog region cold. If it had warmed up, much of the fog would, of course, have evaporated.

Water balance

Another paradox, common to all persistent fogs, is the fact that droplets of water are continually falling to the ground, and yet there is often no diminution in the number of droplets in the air. True, there is always plenty of water vapour about, but this is needed to keep the air saturated. In the London fog the droplets had to fall an average of only 250 ft to reach the ground, a process which took perhaps 6 hr (D). Consequently, then, 800,000 tons of water per day left the fog and, somehow, an equal amount must have replaced it.

Coal contains hydrogen, and its combustion must have added 35,000 tons of water vapour to the air each day, while humans contributed possibly 2000 tons. The remainder, 763,000 tons/day, must have been evolved by evaporation from the ground. The heat coming up from the ground was sufficient to evaporate 18 million tons of water per day; it is reasonable to suppose that 4·3 per cent of this heat was used for the evaporation of water which subsequently replenished the fog.

Smoke balance

Having arrived at a plausible explanation of how fogs can persist in spite of apparently losing water and gaining heat, we can now try our skill on the pollution in the fog.

Smoke particles entered the London fog at a rate of 1000 tons/day, and they must have left it at the same rate, for they maintained a fairly steady equilibrium, averaging 2·2 mg/m^3, or 380 tons in the whole region. It is easy to see that the average smoke particle must have remained free in the air for 380/1000 of a day, about 10 hr.

What ended the life of a smoke particle? They are far too small to fall under their own weight through as much as 250 ft in 10 hr. A rough calculation shows, however, that any smoke particle must collide every 2 min or so with a fog droplet. Could it not, after some 4–10 hr, have stuck to one of these droplets and later settled with it on the ground? There was plenty of dirt on the pavements which might have originally been smoke.

If this view is accepted, it leaves considerable uncertainty over the length of life of a smoke parcticle before it stuck to a fog droplet. For one thing, it is not known what fraction of the fog droplets was capable of entering the volumetric smoke and sulphur dioxide apparatus; i.e. it is not known whether the smoke that was measured, and estimated to be 380 tons, was all free smoke or whether some of it was attached to fog droplets.

Fortunately there is no similar difficulty in connection with sulphur dioxide. Any sulphur dioxide molecule that was attached to a fog droplet would be oxidized, and caught by the filter paper before it could enter the hydrogen peroxide bubbler; thus only free sulphur dioxide was measured.

Sulphur balance

Sulphur dioxide entered the London fog at a rate of 2000 tons/day, and left it at the same rate, maintaining a fairly steady equilibrium of 370 tons in the air. The average free life of a sulphur dioxide molecule was therefore 370/2000 of a day, about $4\frac{1}{2}$ hr.

A few sulphur dioxide molecules dissolved in the water on the ground, vegetation etc., but most of them were removed by the fog droplets whose total surface area, for example, was about 170,000 km^2 compared with only 1160 km^2 of ground. The average sulphur dioxide molecule spent perhaps 0·05 per cent of its time (D) dissolved in fog droplets, moving freely in and out of them. Eventually, after a free life of $4\frac{1}{2}$ hr, during which perhaps 8 sec were spent within droplets, it became oxidized within a droplet, and remained fixed there as sulphuric acid. Six hours later, on an average, the droplet fell to the ground and the sulphuric acid was neutralized by minerals on the ground.

In accordance with the above calculations, the mass of sulphuric acid in equilibrium in the fog must have been 800 tons (D). Its concentration was 4·5 mg/m^3 in the air, and on an average the fog droplets were a 0·4 per cent solution of sulphuric acid.

The possibility of sulphur dioxide becoming attached to smoke particles exists, but a smoke particle can only absorb a small fraction of its weight of sulphur dioxide.

Halogens

Chlorine from coal entered the air in the form of hydrochloric acid at a rate of 140 tons/day. It would be quickly dissolved by the fog droplets, remaining afterwards in the air an average of 6 hr (D). Corresponding to this, its concentration was 0·2 mg/m^3 in the air or 0·02 per cent in the fog droplets. About one-tenth of these amounts of fluorine were also present.

Oxides of carbon

There is no obvious way in which either oxide of carbon can be quickly removed from the atmosphere. It seems probable that the concentrations of both oxides increased steadily throughout the period of fog, and that they dispersed only at the end when the fog itself disappeared.

On this basis, it can be calculated that carbon dioxide reached a maximum concentration in the air of 0·4 per cent by weight, 10 times its natural concentration. The corresponding maximum concentration of carbon monoxide was 150 mg/m^3, or 0·015 per cent by weight.

Ground Level Concentrations

Most of us spend our lives so near the ground that the pollution measured very near ground level is closely equivalent to what we experience. There are perhaps three important factors in calculating ground-level concentrations: calculation of plume centreline concentrations and calculation of spread rates (determined by efflux characteristics and meteorology) and consideration of the time period over which ground-level concentrations are to be observed or calculated.

Instantaneous values are often the highest, but may or may not be injurious to health. Long-term background pollution levels may be of more relevance. To calculate values for a single chimney and then for a multi-stack situation, it is vital to specify the averaging time. The relationship between expected concentrations over different sampling periods is given in Table 31.

TABLE 31. *Time-mean relationships of SO_2 ground-level concentrations*

Sampling period	Mean value of concentration	Maximum concentration (less than 2% of all occasions)
3 min	C/5	1.3 C
1 hr	C/13	C/1·5
1 day	C/60	C/7.5
1 month	C/230	C/45

C is the maximum 3-min concentration.

The maximum ground level concentration occurs at about 10–15 stack heights downstream (see Fig. 114) and the contours of equal concentration are found to be "cigar-shaped" (Fig. 115). Such calculations are based on the classic work by Sir Graham Sutton (1932, 1947, 1953) but are only strictly valid for a steady wind and an homogeneous turbulence field. Modifications have been made to consider the influence of stability (e.g. Pasquill, 1974; Turner, 1970) and nomograms have been produced by, for example, the Tennessee Valley Authority (see, for example, Noll and Miller, 1977), to calculate 1-hr averages for different effective stack heights, emission rates, wind speeds, and atmospheric stability.

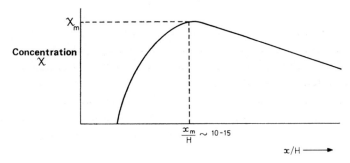

Fig. 114. Ground-level concentrations along the plume axis

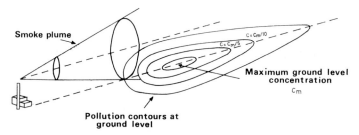

Fig. 115. A pollutant dispersing under conditions of neutral stability

Whenever Sutton's equation is applied to a practical problem, assumptions must be made about the magnitudes of the turbulence coefficients. In most cases, great refinement is probably unnecessary. R.S. Scorer and C. F. Barrett (1961) found a very simple assumption to be useful in estimating the average pollution at ground level from a single chimney. They imagined a cone to be constructed whose apex is the top of the chimney and whose base is a circle on the ground at a radius of 15 chimney heights from the base of the chimney. Between this and a second (inverted) cone standing on its apex at the top of the chimney, they assume the pollution to be spread uniformly. They have constructed a table (Table 32) for the long-term average, at any point on the ground, of pollution from the chimney.

To use the table, one requires to know Q, the mass of pollutant emitted per second, and H, the effective chimney height allowing for any buoyancy of the flue gases. For the table as it stands, all wind directions were taken as equally likely to occur (unlikely to

TABLE 32. *Table for estimating concentrations of pollutants*

Distance from chimney	5H	7·5H	9H	12H	15H*	20H	25H	33H	45H	60H
Long-term average concentration of pollutants in μg m^{-3}†	Negl.	28	56	139	167	153	131	92	56	36

* The maximum concentration occurs at 15H
† All figures to be multiplied by Q/H^2; where Q is in kg/hr and H is in m.

be the case for most observers in Great Britain); an adjusting factor, proportional to the directional wind frequency, should be used for estimating the average concentration in any particular direction from the chimney.

The above considerations are applicable when there is a single stack—a rare occurrence nowadays. More usually the pollutant level at a given location will be influenced by several sources. To discriminate between sources affecting and not affecting the observer the "backwards–forwards" principle can be used. This states that

(i) the expanding cone of pollution from stack 1 will influence all observers between points A and B (Fig. 116) situated on a line, drawn perpendicular to the axis of the plume at a distance d from the stack, and similarly stack 2 influences observers between C and D.

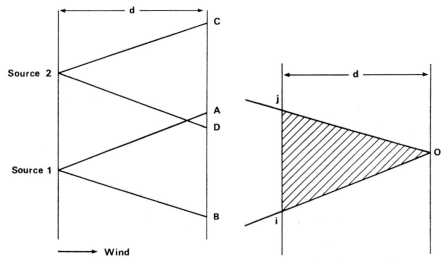

FIG. 116 AND 117. "Backward–forward" principle

Observers between A and D will thus receive pollution from chimneys 1 and 2 and any between. Thus

(ii) an observer at point O (Fig. 117) receives pollution from all sources i to j found by drawing a cone *backwards* towards the chimneys. *All* chimneys within this cone (shaded area on Fig. 117) will affect the observer at O.

Summary

Although there may be other causes yet to be discovered, the more important known causes of change in pollution have now been discussed, and they are recapitulated in Table 33. The rate of emission of pollution is affected by trends and random variations, by yearly, weekly, and daily cycles, and by weather, especially temperature. The quantity of pollution at any particular place is affected by the rate of emission of pollution in an area the size and position of which depends on the direction of wind and other weather conditions; in addition, the quantity of pollution depends on rainfall, wind velocity, and turbulence. The meteorological variables are subject to yearly and daily cycles as well as irregular variations.

TABLE 33. *Causes of change in atmospheric pollution*
The chart should be read downwards, with the words "depends on" inserted at every step

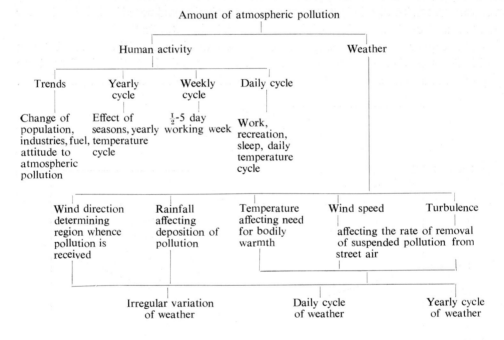

All these causes come into play simultaneously, and they affect different forms of pollution differently. Perhaps it is not surprising that it was necessary to begin this chapter with the remark that the study of atmospheric pollution is an untidy science.

CHAPTER 14

Effects of Pollution

COMPLEX as the relations are between atmospheric pollution and fuel, methods of combustion, human habits, and weather, a scientific study of its effects on everyday life is harder still. This is not because the effects are small, but because they are slow in operation and usually complicated by the effects of other factors. The effect on health is particularly hard to assess. In the general level of health the differences between a poor district of a town and an equally poor rural community are often very considerable, but they are not entirely due to atmospheric pollution. Other factors, such as overcrowding, occupation, and habits play their part, and their consequences are not easily separable, even by statistical methods, from those of pollution. For quite a different reason, the effect of pollution on a normal, healthy individual is also difficult to measure. This is because the human system can compensate to a remarkable degree for departures from perfect health such as would be caused by an unsuitable environment. With careful study however, the harm done by pollution to the sick, the young, and the aged is becoming increasingly well known.

Where the effects of pollution can be scientifically measured they are rather more easy to assess. Over the years, the waste of unburnt fuel, the extra cost of artificial lighting attributable to smoke, the cost of cleaning and replacing smoke-soiled furnishings and clothes, the damage to building materials and paint have been estimated approximately in terms of monetary loss. The effects on metals and vegetation have been tested experimentally. In the present chapter, some instances are quoted of particular effects of atmospheric pollution; where possible, numerical data are given, but it must be recognized that the available information is far short of what is desirable.

Biological Effects

Health

How much harm is done to public and individual health by urban atmospheric pollution is still a matter of opinion, though many authorities consider the total harm to be very serious indeed. Naturally, the effects of atmospheric pollution on the human system have been most closely studied in the severe and fatal cases, and it will be well to begin with the effects of the heaviest concentrations of pollution. The diseases caused by inhaled dusts in mines, works, and factories are now well understood. Under the general heading of *Pneumoconiosis* they include silicosis, a progressive inflammation of the lung tissue which, once it has begun as a reaction to the common substance silica, apparently cannot be arrested; and asbestosis and other forms of reticulinosis, in which particles put out of action many times their volume of lung. Although it is a serious occupational disease, causing the deaths and debilitation of many men in Britain, there is no evidence that pneumoconiosis in any form has ever been contracted by breathing street air.

Chronic bronchitis, in the past termed the English disease, is characterized by a persistent cough and the exuding of muco-pus. The sufferer's breathing is hampered by bronchial congestion and fluid, and the bubbling sound of air through this fluid is used by the doctor in diagnosing the disease.

In a typical year in the 1950's 30,000 people died through bronchitis in England and Wales, and 26 million working days were lost. The annual death-rate was 55–60 per 100,000, compared with 4 for Denmark, 22 for Belgium, and 2 for the United States (see Table 34). The damp foggy climate of Britain was only partly to blame, since there were sharp differences in the incidence of bronchitis between towns and rural areas, between the South and the industrial North, and between England and Northern Ireland. In every comparison the area of heavier pollution came off much the worse, and atmospheric pollution must have been a major cause of chronic bronchitis.

In the early sixties, the death rate rose (71·3 per 100,000 in 1963) but figures for 1968 showed some decrease—see Table 34.

TABLE 34. *Number of deaths per 100,000 of population*

	Early 1950s	1968
England and Wales	55–60	57·5
Denmark	4	7·6
Belgium	22	18·7
U.S.A.	2	3·2
Northern Ireland	—	44·5

Both smoke and sulphur dioxide seem to be involved, and the suggestion is often made that their association together is more dangerous than an equivalent concentration of either separately (synergism). The Bronchitis Research Unit of St. Bartholomew's Hospital, London, has found that the self-observed reactions of a volunteer group of bronchitic patients form a very sensitive index of atmospheric pollution.

In the past two decades, the situation has improved remarkably—there are fewer deaths attributable directly to air pollution and there is a greater understanding of the links (both chemical and biological) between ill health and pollution. However, not all the questions have been answered and disagreement still exists. Particulates greater than about 2 μm in size are unlikely to penetrate the body's biological defences to reach the alveoli of the lungs. Gases such as sulphur dioxide can of course penetrate more effectively, although it is not certain whether high sulphur dioxide concentrations alone are capable of inducing disease—it is highly likely that the cause–effect relationship expressed in Fig. 118 is valid for the susceptible and those already suffering from disease. Distinct thresholds seldom exist so determination of positive/negative responses to selected pollutants is difficult to establish, even statistically.

Lead originating from tetra-ethyl and tetra-methyl lead added as an anti-knock agent to petrol is emitted into the environment and is a great cause for concern, especially in the neighbourhood of large motorway interchanges such as Gravelly Hill near Birming-

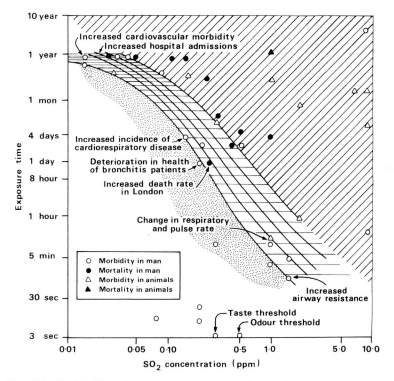

FIG. 118. Health effects due to various exposures of SO_2 (after Williamson, 1973)

ham. Recent medical evidence suggests that lead accumulated in the body can decrease a child's IQ. Other vehicular emissions such as carbon monoxide are also suspect; yet it is evident that the amount of carbon monoxide in the blood of smokers versus non-smokers is far greater than for urban versus rural dwellers. Smoking has been conclusively shown to have the greatest effect on health. Carbon monoxide levels in the blood and incidence of lung cancer are both increased by cigarette smoking. Medical questionnaires today begin by asking whether the interviewee is a heavy smoker; before further investigating exposure to pollution. A recent move has been made by the WHO in the furtherance of public health to ban *all* cigarette advertising.

Threshold Limit Values

"Tolerance levels" of a wide range of pollutants are summarized in the Guidance Note EH 15/77 from the Health and Safety Executive, *Threshold Limit Values for 1977*. This is a revised version of a Technical Data note for 1975 and is based upon values adopted by the American Conference of Governmental Industrial Hygienists at its annual conference in 1977. For most pollutants there is no distinct line between safe and unsafe concentrations and the TLV is only a guideline for safe operation. The values are designed for exposure of workers over an 8-hr period (40-hr week) often in a confined space—acceptable atmospheric concentrations are usually given by the TLV divided by 40. The recommended values are in no way intended to be absolute and are likely to be revised in the light of new knowledge. Certain chemical compounds are best restricted in terms of a ceiling concentration rather than the average concentration described above, and known carcinogens such as blue asbestos, beta-naphthylamine are listed separately.

Smog disasters

Smog disasters are a result of the combination of smoke (or other pollution) and fog. (N.B. The word is now also used to describe the photochemical "soup" present in urban areas as a result of photocatalytic reactions between components of vehicular emissions—see Chapter 10.)

Particulate and gaseous pollution in the open air seldom reaches high enough concentrations to be the obvious cause of death, but when it does there is a major disaster. The first of these was in December 1930, in the Meuse Valley, near Liège, Belgium, occupied by a number of iron and steel works, zinc works, glass works, potteries, lime kilns, electric generating stations, and chemical works including phosphate works. For 5 days a fog persisted and the air within the valley seemed to remain stagnant. Sixty-three people died as a result of atmospheric pollution, the concentration of which was not measured. The victims all suffered from acute irritation of the respiratory system and vomiting, until death came through heart failure. Several hundred other people were severely attacked with respiratory troubles, and many head of cattle had to be slaughtered. According to the *Lancet* **151**, 835, 1946, the substances most likely to have caused the Meuse Valley disaster appear to have been fluorides emitted from phosphate works.

In November 1948 another 5-day fog at Donora, near Pittsburgh, caused the death of nineteen people. The most probable cause was sulphur dioxide from a zinc-smelting works where ores of high sulphur content were roasted. Both these smog incidents occurred in relatively small communities where the death of a number of people within a few days is an obvious disaster. In the big cities, much longer death-rolls were occurring from time to time, without being definitely attributed to smog; but no doubt remained after the calamitous smog in London in 1952.

Mortality attributed to smog

Much careful research has been conducted since the London fog of 5–9 December 1952, in which about 4000 people died above the normal death rate for Greater London. Figure 113, similar to that in the Interim Report of the Beaver Committee, shows how the death-rate rose during the fog period and remained abnormally high for a week afterwards. The highest daily average of sulphur dioxide then observed at Westminster Bridge was 3·7 mg/m^3; in the twenty previous years it has not exceeded 2·2 mg/m^3. Ministry of Health Report No. 95 showed that in one week over 9 times as many deaths from bronchitis were registered as in the previous week. Earlier fogs in 1873, 1880, 1882, 1891, 1892 and 1948 had unmistakably increased the London death rate, but none by nearly so much as in 1952. The Report considered also two fogs in Glasgow in 1909, the Meuse Valley incident of 1930 and that of Donora, Pennsylvania, in 1948. All these fogs were of three or more days duration, in winter, in industrial or densely populated areas where much coal is burnt. There were differences in the pathological reports and the post-mortems, but the concentrations of pollution must have all been rather similar, except that in the Meuse Valley there were additional fluorine compounds from a phosphate works and at Donora there was additional sulphur dioxide from a smelter. In some cases, but not, for instance, in London 1952, the weather was exceptionally cold.

Figures 119 and 120 illustrate some of the work that was done on the measurement of pollution during fog in London by the Fuel Research Station of the then Department of Scientific and Industrial Research. (Clean Air Conference, 1958). Concentrations of sulphur dioxide in parts per 100 million may be multiplied by 2·86 to convert them to milligrammes per 100 m^3.

None of the incidents left incontrovertible proof that one particular form of pollution was the outstanding harmful ingredient. The various official committees agreed that the smog mortality was due to irritation of the respiratory tract of persons already suffering from respiratory or cardiovascular disease. It has been said that the fatalities would have been very much fewer in the absence of sulphur oxides and acids though some medical authorities are not fully satisfied by the evidence at present available. Recent evidence is that the sulphuric acid fumes produce lung damage in guinea pigs similar to that of people who died in the London fog. Further, test animals have been found to be protected from sulphuric acid fumes by the ammonia generated by bacterial action on their own urine and faeces. The show cattle which died in the 1952 London fog had their straw changed very frequently. Sheep and pigs near by, whose lairs were cleaned less often, survived. On the other hand, in the fog of 1875, cattle died in their byres where conditions were probably not clean.

The medical understanding of the disasters is being improved in retrospect by laboratory experiments and studies of records. The programme of the Medical Research Council includes work on the composition of polluted atmosphere, and the effect of exposing animals and humans to each of its constituents; studies concurrently of data on morbidity, mortality, weather and pollution; and studies of respiratory disease in areas of contrasting pollution and in different social environments.

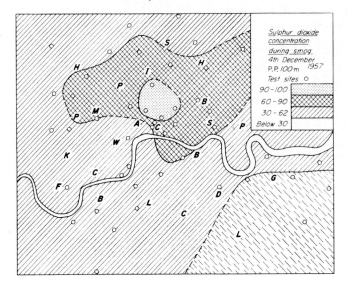

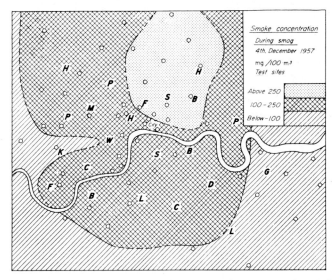

Figs. 119 and 120. Sulphur dioxide concentration and smoke concentration during smog, 4 Dec. 1957

During 3–7 December, 1962, there was a densely polluted fog in London, when the concentration of smoke rose to 1·9 mg/m^3 and that of sulphur dioxide reached 4·1 mg/m^3— comparable with observations in the 1952 fog of 4·4 and 3·8, respectively. The first thing to notice is that the simple comparison 1962/1952 shows an increase in the sulphur dioxide concentration at the same time as a fall in smoke concentration. This indicates that the Clean Air Act and other measures against smoke can be relied on to reduce the extreme concentrations of smoke (of so much concern to the public) as well as the average concentrations which can be more easily studied by the scientists.

The second noteworthy feature of the 1962 smog was also very encouraging. It was reported by Dr. J. A. Scott, Medical Officer of Health for London County Council. Figure 121 shows his main point, that deaths during the 1962 smog were very much fewer that 10 years before. He estimated that the smog killed 340 people in the London Administrative County, comparable to 2000 deaths in the 1952 smog. We can attribute this improvement mainly to the reduction of smoke and associated components of pollu-

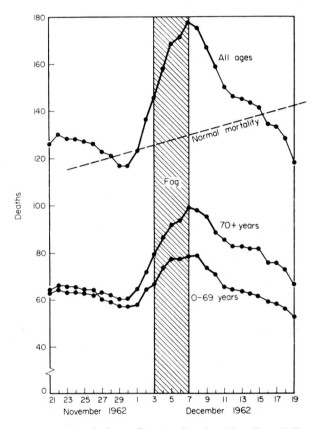

FIG. 121. Deaths from all causes, London, Nov.–Dec. 1962

tion, and also partly to the publicity given since 1952 to the harmful effects of smog. Many people with chest complaints just did not go out of doors during the worst days.

In 1953 and 1954 two reports were presented by the Beaver Committee. They recommended that during fogs the aged and infirm should remain indoors and rest as much as possible; or to wear a "smog mask" when out of doors. Over 20 years later, such ideas are, happily, outdated as Britain has been little troubled with smogs for more than 15 years. No longer does the Meteorological Office issue "smog warnings" on a regular basis. (In fact it is perhaps reassuring to hear pollen counts instead.) Increased pollutant concentrations do not last long; although there is little room for complacency. Success with smoke must be matched with completion of the reduction of sulphur dioxide and other pollutants which can cause degradation of ourselves and our environment or simply be a non-toxic nuisance (e.g. powerful malodours).

Effects on animals

Reference has already been made to mortality among animals in the London fogs.

Animals may be expected in general to suffer similarly to human beings. Cattle are found to be less resistant than sheep and pigs. Sheep and cattle are particularly sensitive to fluorine. In districts where emissions occur, fluorine tends to accumulate in the grass; the teeth of grazing animals have been known to deteriorate so much through fluorosis that the animals were unable to feed, and cases have been known of cattle deaths resulting from poisoning by molybdenum probably originating from a molybdenum sulphide roaster.

Effects on vegetation

Plants are more sensitive than animals to atmospheric smoke and sulphur dioxide, though this is not true of all forms of pollution. For instance, grass assimilates an appreciable proportion of fluorine without harm. Smoke is particularly harmful to plants, and it has been showed that radishes and other plants lose from half to nine-tenths of their growth in a polluted atmosphere, if they survive at all; that soil loses its stock of accessory plant foods very rapidly under the action of polluted rain; and that the amount of pollution in any particular district may be inferred with remarkable accuracy from the type and condition of the vegetation.

Lichens have been well used as indicators of sulphur dioxide pollution, although in an ameliorated environment growth appears to be more strongly forced by environmental and meteorological factors. Moss bags have been used with good effect in estimating airborne concentrations of heavy metals.

Plants are most susceptible when environmental factors encourage their stomata to open (e.g. high humidity, light). It is possible to (approximately) rank species by their general susceptibility.

On the effects of sulphur dioxide, research in Canada revealed that lucerne was injured by as little as 0·3 parts per million and barley by 0·8 parts per million; in concentrations of over 1·0 parts per million, increasing proportions of the foliage and fruit were destroyed. Experiments by Professor Mansfield at the University of Lancaster with various grasses have demonstrated the stunting effect on growth of both sulphur dioxide and nitrogen oxides (NO_x). Combination of these two pollutants has a strong synergistic effect where the total damage of the mixture is greater than the sum of the damage of the two pollutants administered separately.

In areas where pollutant levels (especially sulphur dioxide) are still high, immunity can be developed. Experiments in Liverpool have shown that there is a strain of rye grass which has adjusted to the relatively enhanced levels of sulphur dioxide and now thrives.

Physico-chemical Effects

Insulators

Over the years the electricity authorities have suffered appreciable energy losses through atmospheric pollution on the surface of insulators used in suspending high voltage transmission lines. In humid weather the dirty insulators cause "flashovers" which interrupt the service, and apply undesirable stresses to the transmission system.

Metals

Experiments have shown how rapidly, and in what circumstances, the different metals corrode. The rate at which iron rusts is greatly accelerated by traces of sulphur dioxide in the air, and by particles of smoke or ash; in the presence of these forms of pollution (but not otherwise) there is a very great increase in the rate of rusting when the relative humidity of the air rises above about 80 per cent. Rusting is most rapid when all three factors, humidity, sulphur dioxide and particulates, are present together. For example, a specimen of iron carrying particles of carbon was exposed to a humid atmosphere containing traces of sulphur dioxide; it showed approximately a hundred times as much corrosion as an identical specimen, similarly exposed in the absence of sulphur dioxide. Another specimen of iron was exposed in a muslin cage which permitted air and sulphur dioxide to reach it, but excluded particulates. Its rate of rusting was negligible compared with that of a similar specimen which was not inside the cage.

Other metals are attacked in different ways by atmospheric pollution. Nickel catalytically oxidizes atmospheric sulphur dioxide and accumulates a film of sulphuric acid; this reacts with the metal to form basic nickel sulphate. Chromium behaves similarly, and chromium plate will not deteriorate if it is frequently washed. Zinc slowly and steadily corrodes in a polluted atmosphere, forming basic zinc sulphate. Copper tarnishes if it is kept dry indoors, forming copper oxide and copper sulphide; but out of doors it forms a "patina" of basic copper sulphate, and this protects it from further corrosion. Silver tarnishes like copper, and it is often worth while for silver articles to be rhodium-plated as a temporary protection, at a cost of about 10 per cent of their selling price. Aluminium and its alloys form a protective film which resists the action of atmospheric pollution.

Materials

Wool, cotton, and leather materials are rotted in polluted atmospheres as a result of the absorption of sulphur dioxide and its oxidation to sulphuric acid, which attacks the proteins in the materials. In a humid atmosphere containing 10 parts per million of sulphur dioxides, specimens, of leather will rot in 6 weeks, unless they are protected by rubbing with a solution of potassium lactate or certain other organic salts. The deterioration is much slower in ordinary atmospheres, but is important, for instance, in leather book bindings, particularly in urban libraries.

Works of art can be most seriously affected by pollution. Smoke is harmful to paintings, particularly on wood; also to textiles, leather and wax-film material. Sulphur dioxide affects oil- and resin-film material, leather, paintings (particularly on mud wall, paper, or wood) and vellum. The slightest deterioration of any work of art is naturally a matter for concern.

Building materials are both corroded and disfigured by atmospheric pollution (Figs. 122, 123). Smoke particles stick on every surface of stone, brick, paint, or glass, whether

FIG. 122. Weathering at Chelsea Hospital, London

FIG. 123. Unrestored and restored portions of stonework, Henry VII Chapel, Westminster Abbey

horizontal or vertical, forming a film of soot which may or may not be partly removed by rain. The sandstone buildings of cities in the Midlands and the north of England used to be blackened soon after they were built, the details of the architect's design disappearing into a dingy uniformity. Limestone buildings go into half-mourning, their more exposed surfaces being kept clean by rain. Some public buildings of polished stone and limestone are regularly cleaned by hosing.

Indeed it is only since smoke levels have been drastically reduced that local authorities have felt it worthwhile to clean their buildings by sandblasting, vastly improving the visual impact of these magnificent edifices. It is interesting to remember that after 100 years on the Thames Embankment, Cleopatra's Needle has been eroded by an amount equivalent to that resulting from 3000 years exposure in Egypt.

Although some stones are relatively immune, certain sandstones, limestone, roofing slates, mortar and reconstructed stone are prone to attack by sulphur dioxide and the acid constituents of atmospheric pollution. Any building material containing carbonates suffers damage since the insoluble carbonates are converted by polluted rainwater into soluble sulphates or chlorides; the stone disintegrates underneath the surface and ultimately a large flake falls off. An authoritative estimate in 1930 put the cost of the damage by pollution to building property in the United Kingdom at £2–2$\frac{1}{2}$ million per year.

Fog, visibility and sunlight

Visibility may be defined as the maximum distance at which scenery can be distinguished, in daylight, against the sky as a background. It varies in different conditions from a few metres to over a hundred kilometres, and the poorest visibility is always caused by fog, any smoke particles present being relatively unimportant. In the absence of water droplets, however, smoke used to be a frequent cause of poor visibility, and a real danger to aircraft which require a visibility of about a kilometre when landing.

The following two paragraphs describe the damaging effect reduced visibility must have had on the population of Britain 20 years ago.

> The part played by smoke particles in reducing visibility may be seen by considering the records of visibility, made by meteorological observers on days when there is no possibility of fog. On such days, the average visibility-distance in winter from Kew Observatory is 20 per cent less than in the country near London. The difference must be due to extra smoke in the air around Kew, and it should be remembered that Kew has not by any means the smokiest nor the country round London the cleanest air in Britain. At Leicester it was found that when the concentration of smoke was doubled the visibility-distance was halved; and it was deduced that a ton of smoke (suitably placed) is sufficient to blot from overhead view nearly a square mile of country.
>
> On smoky days in winter, townspeople often need artificial indoor lighting an hour or more before sunset, and may even need it all day, whereas in summer they can read by daylight until after sunset. The setting sun should always be equally bright, but in winter in many districts, and throughout the year in some, it is frequently no more than a dull patch of red. In this era of cheap and easy artificial lighting, it might seem that the loss of daylight causes no hardship other than an occasional unexpected lighting-load on the electritcity generating stations. But the human body requires natural light and ultraviolet radiation to help in its perpetual struggle against micro-organisms and to produce vitamin D, and a considerable proportion of urban ill-health is attributed to lack of "sunshine". Insufficient visible and ultraviolet radiation is believed by the medical profession to give rise to general ill-health, and especially to a high incidence of tuberculosis and rickets in children. From measurements in a number of British cities, by comparing the daylight received at points outside and inside the city, it is estimated that from 25 to 55 per cent is lost through smoke alone, in the five winter months November to March. The losses of ultraviolet radiation are similar. On the gloomier winter days nine-tenths of all radiation is lost.

Nowadays pollutant concentrations are rarely responsible, by themselves, for artificial lighting during the day as described above. Although in city centres the number of sunshine hours is consistently less than in the more rural surroundings, the number of hours of sunshine has increased by a large factor over the last 25 years. (In the case of

Manchester, the winter sunshine over the period 1950–1969 increased by almost 100 per cent—see Fig. 124.) Winter sunshine at Kew and London Weather Centre are now almost identical.

A recent survey was carried out by the Medical Research Council, St. Bartholomew's Hospital Medical College, into the effects on health of migrants between London and Crawley in comparison with life time residents of Crawley. Little difference was found, the major effect on health being cigarette smoking and a history of childhood respiratory illness. All subjects belong to one of two groups—those born in 1952 or in 1957.

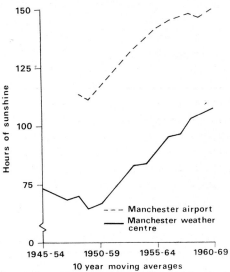

FIG. 124. Number of hours of sunshine during Nov.–Jan. at Manchester Airport and Weather Centre 1945–1970 (after Wood *et al.*, 1974)

A recent survey in an industrial U.S. town (in Connecticut) of respiratory health revealed no significant differences in comparison with a more rural control group (Bouhuys *et al.*, 1978).

The Cost of Pollution

Although there is no way of evaluating the effect on health, many of the harmful effects of atmospheric pollution which have just been discussed can be measured approximately in terms of money. The following estimates have been made of the cost of atmospheric pollution from all assessable causes, including the loss of unburnt fuel in smoke, but excluding loss of health: Pittsburgh, Pennsylvania, 1912, 20 dollars or £4 per head per annum; London, 1924, £1·2 per head per annum; Manchester, 1919, well over £1 per head per annum; Manchester, 1924, £1·5 per head per annum; in the whole of Great Britain, £40–50 million per annum in 1924, and about £100 million per annum in 1947, £100 to 150 million per annum in 1953, and £250 million per annum in 1954 (Beaver Report). A report in 1972 gave the direct annual cost of damage caused by air pollution as approximately £410 million for the U.K. as a whole. An additional estimate for costs associated with social health and amenity costs results in an estimate of almost £1200 million (or £21 per head of population). (The escalating costs are, of course, partly due to inflation). In May 1979 the London Boroughs Association estimated that they were spending over one million pounds each year to repair damage to buildings in central London caused by "acid rain".

Certain individual items on the bill of costs are of special interest. Household washing is more troublesome in industrial cities than in country towns, not only because of the difficulty of drying white articles without getting them dirty from atmospheric pollution, but because on an average there are more articles to be washed. The extra cost in 1919 of the materials and fuel used for household washing in Manchester, compared with Harrogate, was 3p per week or £1·6 per year per household. The extra time required for the household wash in Manchester was about 1 hr per week.

The value, as a fuel, of the 2·4 million tonnes of smoke made each year in Britain was then estimated at about £12 million but this is not all. An example was given in Chapter 7, Fig. 33, of smoke that carried with it from 3 to 5 times its own weight of invisible combustible matter. If it is legitimate to generalize from this particular instance, the invisible combustible matter that accompanies smoke must have been worth a further £40 million, even after a generous allowance had been made for the combustible matter which is necessarily wasted under the optimum conditions of combustion in a furnace.

Conclusion

The harmful effects of atmospheric pollution are so widespread and varied that they are difficult to summarize. There is no doubt whatever that atmospheric pollution in the concentrations in which it has been allowed to occur, particularly in urban areas, caused damage to property and made living conditions generally less pleasant, nor that concentrations sometimes occurred which could be held directly responsible for immediate serious damage to plants and even loss of life. Although the damage caused by pollution has been reduced during recent years, human health, animal and plant growth and survival are still affected detrimentally. New causes for concern have arisen; increased road traffic has enhanced NO_x concentrations which together with increased sunshine (caused itself by a decrease in smoke levels) has increased the incidence of photochemical smogs; supersonic planes chlorofluorocarbons and the ozone layer in the stratosphere; increasing carbon dioxide may disrupt the stability of the global climate. Such a shift in emphasis should not reduce our concern over smoke, sulphur dioxide and respirable dust and make us oblivious of local pollutant levels which are, in general, no longer as immediately obvious to the "man-in-the-street".

CHAPTER 15

Prevention of Atmospheric Pollution

IN THE different states and cities of the world there are widely differing attitudes to pollution. Some areas are either fortunate that they have little pollution to contend with, or unfortunate that other problems, of health or economics, have to be placed higher on the list. However, in many places the problems of pollution are both important and urgent. The purpose of the present chapter is to summarize the practical ways in which atmospheric pollution can be reduced. In an industrialized country and in the absence of legislation, the degree of pollutant prevention must be determined by the cost to the industry balanced against benefits to the general public, a decision largely forced by public concern about their health and their environment together with the employer's concern about the health of its employees. Sometimes the addition of air-pollution-control devices may be of direct benefit to the industry permitting saleable waste products to be collected. In Britain the legislation of the Alkali Acts introduced the phrase "best practicable means" (a phrase reinforced by Clean Air legislation) in order to minimize pollution whilst recognizing that its total elimination would often result in the inability of the industry to function economically at all. Advice on "best practicable means" comes from the Alkali Inspectorate and the local authority Environmental Health Officers.

Prevention of Smoke

It is no longer debated whether smoke elimination is justified on purely economic grounds.

It is technically not difficult to burn coal smokelessly in medium or large shell-type or water-tube boilers; and although instruments and labour of a relatively high standard are required, in many instances a reduction of smoke is likely to be achieved at a negative cost to the management.

Categorical statements are notoriously easy to refute, but this one is almost universally true: raw coal cannot be burnt in domestic grates, boilers or furnaces, without making smoke. Resort should be made to alternative fuels. Domestic smokeless fuels will show a saving even if they cost up to 30 per cent more than coal. Industrial smokeless fuels are often advantageous in themselves; for instance, most metallurgical and other heat-treatment processes are more efficient and more controllable when fired by gas or oil, instead of coal.

The dominance of steam locomotives on British Rail (main line) was gradually lessened by electrification programmes and the, supposedly temporary, introduction of diesel locomotives. In Britain no new steam locomotives were put into service after 1960 and the era ended when 70013 *Oliver Cromwell* made its last trip in 1968*.

The problem of eliminating domestic smoke is a complex one, because it is essential to convert coal-burning appliances in old houses, as well as to see that new dwellings have their appliances designed for smokeless fuels. The numerous ways of achieving these ends were discussed in Chapter 9. In Britain the Domestic Solid Fuel Appliance Approval Scheme, the Gas Council and the Electricity Boards issue lists of domestic appliances which are revised periodically to ensure that householders and others may obtain up-to-date information regarding makes and types of appliances which have satisfied performance and other tests.

There can be no doubt that the atmosphere of Britain is becoming very different as bituminous coal ceases to be burnt in private houses and as smokeless operation of industrial boilers becomes the "norm". In some districts, however, smoke from special industries such as pottery, metallurgical, or glass works is a major problem, and must be dealt with by special methods, some of which were outlined in Chapter 8.

Prevention of Ash and Grit

The type of pollution most apparent to the eye depends upon the point of view. People seeing an urban district from a neighbouring hill or from the air always notice the increased particulate load most easily because of the great power of absorbing and scattering light. If to people living in it, however, a district has the reputation for being heavily polluted, smoke is not usually the main cause (e.g. near quarries and cement works and newly opened open-cast mines). Unless there are chemical fumes with pungent smells the cause is nearly always traceable to a high rate of deposition of ash or grit; these are to them the most obvious form of pollution.

The total damage done by grit is probably only a few per cent of that caused by all forms of atmospheric pollution, but the damage occurs chiefly within a short distance of the source of the grit, and people who have cause to complain can usually indicate with some certainty the place of origin of the particles. Partly for this reason, perhaps, the immediate causes of grit emission have been closely studied and the development of industrial grit arresters has proceeded more rapidly than that of all other devices for reducing atmospheric pollution.

Particles of ash and grit are emitted from chimneys if two conditions are fulfilled: if particles are available to mix with the gases in the combustion chamber, and if the velocities of the gases in the flues and the stack are sufficient to carry the particles away. The

---

* Footnote to this: 1978/9 saw the reintroduction by British Rail, on a very limited scale, of special passenger trips pulled by famous locos of the past, notably 4472 *Flying Scotsman*.

emission may be prevented or much reduced, therefore, by selection of fuel, by design and operation of the fuel-burning equipment and by implementation of efficient control devices.

Selection of fuel

The washing and grading of coal at the collieries is clearly a useful way of reducing emission. In washing, much of the mineral matter, including a proportion of the finer particles, is removed. In grading, the coal dust is separated from the other sizes and can be burnt in specialized equipment. Particle emission is also reduced if the coal cakes on the fuel bed, when particles fuse together and are prevented from escaping. Where fine slack coal is burnt on chain-grate stokers, it is common practice to wet the fuel on its way to the grate with about 10 per cent of its weight of water or steam; the purpose of the moisture is to hold the particles together until fusion begins.

The solid fuels, then, which are least likely to produce excessive emission of ash and grit are washed and graded coal and good caking coal. Coke and free-burning coal, especially when they contain a high proportion of fine breeze or dust, are liable to cause fly ash and grit. Unfortunately one of the products of a coal-washing plant is "middlings" (or in some cases a slurry) which contains some of the coal mixed with a high proportion of mineral matter.

The worst potential source of fly ash and grit is pulverized fuel, not only because it is entirely particulate, but because it is usually burnt while in suspension in the combustion chamber. All particles of mineral matter in it, together with the ash residue from each fragment of fuel and, usually, some partly burnt particles of fuel, are free to escape into the flues and the atmosphere, unless they are specially prevented.

Design and operation of furnace

Fuel is too valuable to be thrown away, and we must consider the next means of defence against emissions of ash and grit—in the furnace and flues. To begin with, in all boilers having grates, if primary air passes steadily and evenly through the fuel bed, the minimum number of particles will be lifted. If the fuel bed is too thick the conditions

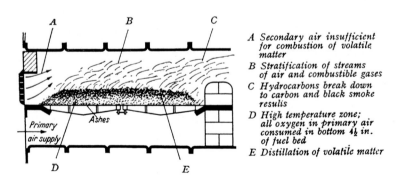

Fig. 125. Effects of a thick fuel bed

shown in Fig. 125 are reached. When a boiler is being forced beyond its rated capacity, the rate of fuel burning is increased by admitting more air than is required for normal combustion. There is a faster air flow through and over the fuel bed and more particles are lifted, including fuel particles which might have had time to burn in the combustion chamber, but which are now swept into the flues without being burnt. In such cases the grit emitted contains a high proportion of unburnt and partly burnt fuel. These considerations apply to many other furnaces besides boilers, but they are especially applicable to large stoker-fired installations. In times of high industrial output, many chain-grate and retort-stoker installations are being forced, with the result that excessive quantities of "smoke" are emitted. The "smoke" may not be black enough to contravene the regulations, but it consists of minute spheres of coke and ash which are responsible for numerous minor eye injuries, and must therefore be eliminated before the waste gases are vented to the atmosphere.

The introduction of fluidized bed combustion (see Chapter 7) will obviate many of the above difficulties by retaining most of the pollutants within the bed.

Particulate removal

When the emission of flue dust cannot be prevented by other means, special devices for removal of particulates must be employed. These can be one of four basic types: mechanical collectors, fabric filters, scrubbers and electrostatic precipitators—choice will depend on size and amount of dust to be removed and on economic considerations. It may sometimes be desirable to use two types in series. The larger a particle is, the easier is its removal by mechanical means; and the largest particles of all, over about 0·1 mm (100 μm) in diameter, can, for example, be allowed to settle under gravity.

Grit or dust can be graded according to size by placing a sample on top of a number of wire mesh sieves which are shaken mechanically. The dust caught on each sieve is weighed and expressed as a percentage of the total weight of the sample. Results are usually expressed with reference to British Standard mesh numbers, or more commonly, to the corresponding aperture sizes in microns as shown in Table 35.

TABLE 35

B. S. Mesh No.	Aperture in microns
60	250
72	212
85	180
100	150
120	125
150	106
170	90
200	75
240	63
300	53

A gravity settling chamber (see Fig. 126) allows the dust to be retained for a sufficient length of time for large particles to separate out under gravity. This can be assisted by the introduction of baffles. Operation costs are low but the chamber is bulky and relatively inefficient (50–60 per cent). For smaller particles an artificial acceleration can be produced by whirling the dust-laden gas round the inner wall of a cylinder. A cyclone dust-separator, which is based on this principle, is illustrated in Fig. 127. The gas enters

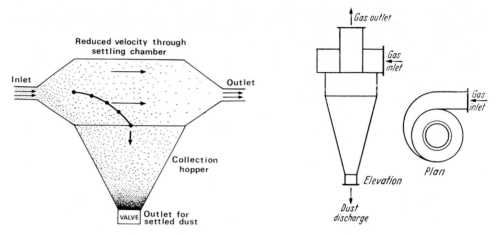

FIG. 126. Gravity settling chamber

FIG. 127. Cyclone dust separator

at high speed tangentially at the top of the cylinder, and takes a course in tighter and tighter spirals to the bottom, where it passes up a central tube. The particles are thrown centrifugally to the walls of the vessel, and find their way to the bottom: they are discharged through an air lock if, as is usual, the gases within the cyclone are at less than atmospheric pressure. Ultrasonic agglomeration may be used as a pre-treatment to increase efficiencies. (This is sometimes regarded as a fifth basic class of collector).

A second method of removal is by use of fabric filters. Dust is retained mainly by interception, electrostatic attraction and inertial impaction (see Fig. 128) on the elements of the filter. The filter bags are often replaced when dirty or may be cleaned *in situ*. The recently developed high efficiency (HEPA) filter is similar but has a more open structure in which interception is the main removal mechanism. The filter material is often pleated

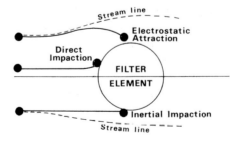

FIG. 128. Methods of capture of dust particles by fabric filter element

paper to increase the area and decrease the air flow, ensuring high efficiency of collection (>99 per cent in most size ranges).

Particulates can also be removed by wet scrubbing, in which a spray of liquid is brought into contact with the particles which then "rain out" — a technique also used for removal of waste gases (see next section) or where particles are forced to impinge on a wet surface. Figure 129 shows an impingement plate scrubber in which the particulates are collected by impingement plates set at various levels in the scrubber.

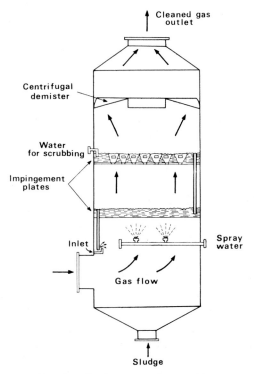

FIG. 129. Impingement plate scrubber

Where it is important to remove a high proportion of particles smaller than 20 μm, the most widely used method is electrostatic precipitation. Figure 130 shows diagrammatically a typical electrostatic precipitator. The flue gases pass in stream-line flow through earthed vertical tubes in parallel, in each of which is a central wire charged to about $-40,000$ V. Ionized gas molecules, travelling at high velocity between wire and tube, collide with the particles in the gas stream and these, too, become electrically charged. Those that are negatively charged are driven to the earthed tubes where they stick. When the tubes are jolted by the vibrator rapping gear the particles, now coagulated into larger aggregates, escape against the flow of the gas stream while keeping near the walls of the tubes, into hoppers. Those particles which are at first positively charged are driven to the central wires from which they are released by the rapping of the wires. They have now the same charges as the wires, namely $-40,000$ V, and they immediately join the other negatively charged particles in their migration to the earthed tubes.

14*

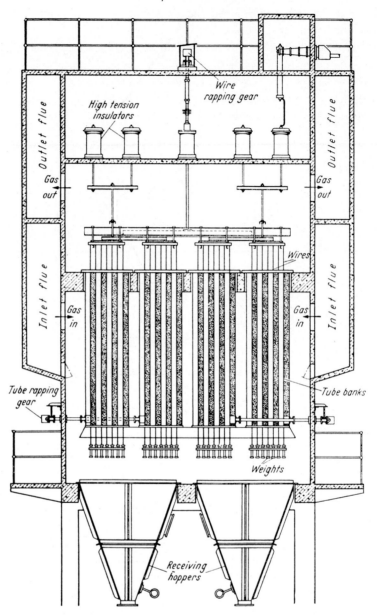

Fig. 130. Electrostatic precipitator with vibrator rapping gear

Electrostatic precipitators are capable in practice of removing 95–99 per cent of the weight of dust from flue gases; they are rather more effective for removing mineral dust than carbonaceous dust such as particles of partly burnt coal.

One serious problem with any partial-removal plant is the disposal of the collected dust, which may amount to many thousands of tonnes each week at a single generating station. Disused quarries and ponds in the neighbourhood can be filled, but precautions must be taken against the dust becoming wind-borne in dry weather; for example, it can be covered with a layer of soil in which vegetation can take root. One of the few cases of a dust storm in England occurred when several acres of dust "lagoons" became so dry that the particles were lifted by the wind.

Air conditioning

In areas where the emission of pollution is particularly high it may be necessary to purify the air before it enters certain rooms: e.g. in hospitals, rooms housing mainframe computers. The task is relatively easy if in any case the air requires heating, cooling, or a change in its humidity, before it is ideal for human consumption. It is, however, a different problem from that of removing dust from flue gases: (1) the quantities of air to be treated are usually not so large, (2) since the air is to be breathed, it cannot be contaminated with poisonous substances, (3) the particles are usually smaller in size and (4) their concentration is much less than in flue gases. Frequently it is possible to use filters of cotton wool or other material without having to renew the filters too often. Electrostatic precipitation is also used, but with reduced voltages, since it is important to prevent the electric discharge from producing ozone and oxides of nitrogen in objectionable quantities.

Prevention of Sulphur Dioxide

At the time of writing of the third edition (1962) no local or national authority had given itself the task of systematically controlling the emission of sulphur dioxide within a region. There were cases of individual electricity generating stations and zinc sulphide smelters where the emission of sulphur dioxide was severely restricted by agreement with the interested authority, but there was no counterpart for sulphur dioxide of a smoke-control area or a smokeless city.

In comparison with smoke and dust, sulphur dioxide is a form of atmospheric pollution which is difficult to prevent, although there are simple ways of appreciably reducing the amounts emitted into the air. To begin with, therefore, it is advisable to have an idea of the cost of the damage attributable to sulphur dioxide. Apart from its effects upon health, in 1962 it was probably costing Britain somewhere between £80 and £200 million per year for its damage to metals, wool, cotton, leather, paint and building materials; that is £16 to £40 per tonne of sulphur dioxide emitted, or 50p–£1·25 per tonne of coal burnt. In densely populated areas the cost was much higher than the average, so it is reasonable to have in mind a figure of at least £1·25 per tonne of coal burnt, when considering the economics of reducing the emission of sulphur dioxide in towns.

The methods available for preventing or reducing damage by sulphur dioxide fall into four classes of which the first two are closely connected:

1. Using relatively sulphur-free fuels, such as wood, natural gas or methane, petrol (gasoline), paraffin (kerosene) and gas oil. To these may be added electric power from hydro-, wind-, solar- or nuclear generating stations.
2. Removing sulphur from fuels before burning.

3. Removing sulphur dioxide from the products of combustion.
4. Emitting flue gases into a region of the atmosphere where they are relatively harmless, by using tall enough stacks, or remotely placed power stations, having undertaken a suitable investigation of the prevailing meteorological parameters.

Removal of sulphur from fuel

The sulphur-poor fuels listed under (1) may of course serve the needs of some isolated communities, but the amounts available are not large, and their wholesale use would badly disrupt the economy of any industrialized country. What are the hopes of adding to them significant quantities of coal or fuel oil from which the sulphur has been artificially removed?

Some of the sulphur in coal is removed at the colliery during hand-picking and washing. Moreover, the consumer who burns washed coal needs perhaps 10 per cent less in total weight and the weight of sulphur is correspondingly reduced. This usually turns out to be the major item in the saving of sulphur dioxide and, on an average, perhaps a total reduction of 15 per cent is achieved by burning washed coal. The economics of each colliery determine how much coal shall be washed.

It has been mentioned that the material removed from coal at collieries is often highly combustible, and that some of it is burnt in the colliery boilers. Some also is thrown on to spoilbanks where it may catch fire spontaneously. In either case, the neighbourhood of the colliery is likely to suffer from sulphur dioxide, and the pollution of towns will have been relieved only at the expense of the colliery districts.

A second way of preventing sulphur in coal from being burnt is to prepare coal gas (a process no longer in general use in Britain) in which much of the sulphur was converted to hydrogen sulphide and was removed in the purifiers. However, carbonization is a most uneconomical way to remove sulphur, and was a fortuitous "by-product" of coal gas production. There are various processes for the complete gasification of coal, with sulphur removal, but they all result in relatively expensive fuels.

What is the prospect of removing sulphur from oil fuels? In 1959 probably about 12 per cent of the sulphur in the oil delivered to Britain was chemically converted to useful products or to non-gaseous wastes, and thus did not contribute to atmospheric pollution. The remaining 88 per cent was presumably burnt and emitted into the atmosphere.

More recent chemical advances make it possible to remove up to 99 per cent of the sulphur but in most industries the amount removed is still determined by legal and economic factors.

The removal of sulphur from the bulk of fuel oils may be commercially unattractive, but it is not impossible. In the various involved mixtures of hydrocarbons which we burn as fuel oil, some of the sulphur is attached to ring compounds which are particularly hard to dissociate. The only way to remove sulphur entirely would probably involve cracking all the fuel oil into lighter oils—a course which the oil companies serving industrial countries would hardly contemplate. However, it must be possible to find catalytic processes whereby over half the sulphur is removed while leaving the fuel oil essentially unchanged.

Removing sulphur dioxide from flue gases

In coal, about one atom in 250 is a sulphur atom; in fuel oil, about one in 200. Put in this way, the task of extracting the sulphur from fuel looks far from easy. It looks harder still to extract sulphur dioxide from the flue gases after the fuel has been burnt. We may now talk of molecules instead of atoms, and when coal is burnt in air, only about one molecule in 900 is sulphur dioxide; in the case of a fuel oil, the corresponding figure is about one in 500.

The main cause of this further dilution of the sulphur is, of course, the nitrogen in the air, playing no part in the combustion but about 4 times as abundant as the oxygen. We can also see that a flue-gas scrubbing plant must handle about 15 times the weight of material dealt with by a fuel sulphur-extraction plant of the same capacity. Moreover, gases are about one-thousandth of the density of the corresponding solids or liquids, so the *volume* of material handled by the gas-extraction plant must be about 15,000 times greater.

The removal of sulphur dioxide (and indeed any other gaseous waste) from flue gases is undertaken in one of four basic ways: by absorption, adsorption, condensation or conversion. *Absorption* can occur with or without a chemical reaction in a wet scrubber. The liquid scrubbing medium must be selected with the gas in mind, e.g. sodium hydroxide removes mercaptans, diethanolamine removes hydrogen sulphide—many other gaseous pollutants are scrubbed successfully by water. In *adsorption* the gas is retained on the surface or in pores—again with or without a chemical reaction. Activated carbon adsorbs light hydrocarbons, silica gel adsorbs water vapour. The efficiency of this process

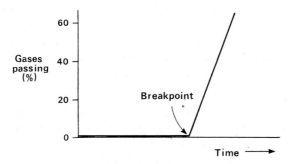

FIG. 131. Breakpoint graph for adsorption

is very nearly 100 per cent until a breakpoint occurs (see Fig. 131), after which regeneration is needed.

Recovery of economically valuable waste gases or corrosive gases can often be best undertaken by *condensation* either by a reduction in temperature or, more rarely, by an increase in pressure. Condensation is also useful in reducing the volume of an effluent gas mixture prior to passage through further control equipment; thus avoiding the need for an excess capacity in this latter apparatus. Two basic types of condenser are used: where the coolant and condensate mix (contact condenser) and where heat transfer is across a solid medium, e.g. a shell or tube type (surface condenser). Condensers are often used in petroleum refining, ammonia and chlorine manufacture and in dry cleaning.

Conversion to none or less pollutant gases is widely used especially for oxidation of combustibles to carbon dioxide and water. Hydrogen sulphide is often removed in this way by conversion to the less toxic sulphur dioxide. Direct and catalytic afterburners are used for many stationary sources and have also been developed for implementation in car exhaust systems.

In Britain removal of sulphur dioxide from power-station waste gas was first implemented at several locations in London: Battersea Power Station used River Thames water to scrub the gases. In 1954 the same process was installed at the oil-burning Bankside Power Station. Fulham Power Station was built in 1935 with packed column scrubbing equipment.

Although a scrubbed effluent forms a plume containing water vapour which rises more rapidly than a "dry plume" as the water condenses and releases latent heat (see Chapter 13), the process of scrubbing tends to decrease the temperature of the effluent. The overall effect of installing wet scrubbers is thus to replace a high rising effluent containing sulphur dioxide by a predominantly water-laden plume, which may rise little or even descend to ground level locally.

It has been stated that the difficulty of preventing the emission of sulphur dioxide from the combustion of fuel is much increased by the low concentrations of the gas in the flues. In smelters where sulphur dioxide is produced from the oxidation of sulphides of zinc and other metals the concentrations are much greater and there is more scope for the chemist. At Traill, British Columbia, where about 800 tonnes/day of sulphur dioxide are produced, equipment is available for flue-gas washing and for producing ammonium sulphate and sulphur as by-products. This equipment cannot be operated continuously, however, and an arrangement has been made for sulphur dioxide to be emitted to the atmosphere only in such amounts, and on such occasions, as to produce less than a definite maximum concentration of the gas at ground level. The direction and type of wind flow is complicated by the fact that the plant is in a deep valley. The services of a resident meteorologist are necessary to take observations and to ensure that the plant engineers have adequate warning when a change of weather is imminent, particularly any change which will affect the flow of air down the valley and across the frontier into the United States.

Chimney height

The above processes remove sulphur dioxide and other waste gases from the flue gas. Although efficiencies are high they are seldom as high as 100 per cent and so some sulphur dioxide is left in the gases which enter the stack. Chimneys were originally designed to supply a draught and tall chimneys were introduced not to eliminate pollutants but to discharge them into the atmosphere at sufficient height that (so it was hoped) they would have adequate time to disperse before reaching ground level. The diffusion of gases down to ground level was discussed in Chapter 13. It has been calculated that (1) the maximum concentration at ground level occurs downwind at a distance, in average meteorological conditions, of 10 or more chimney heights from the base of the chimney (unless the chimney is less than about 2·5 times the height of surrounding buildings, in which case eddy currents set up by the wind in blowing past the buildings may engulf the flue gases and bring them quickly to ground level), and (2) the maximum concentration itself varies inversely as the square of the effective chimney height.

Using a simple formulation assuming normal weather conditions, this can be expressed by

$$P_{max} = 2.71 \ E/uH^2$$

where P_{max} (mg/m³) is the maximum ground-level concentration of pollution due to a particular chimney,

E (kg/day) is the rate of emission of pollution from the chimney,
u (ms⁻¹) is the wind speed,
H (m) is the effective chimney height.

The formula allows for the initial rise of the flue gases above the top of the chimney (by the method of Bosanquet) and is essentially the same as the formula given in Appendix VI of the Beaver Report. The value of P_{max} is believed to be reasonably correct in a wide range of weather conditions, but the distance downwind from the chimney at which the maximum concentration will be found is variable.

The diffusion of chimney gases is strongly dependent on the degree of turbulence, as was pointed out in Chapter 13. When turbulence is strong, diffusion is rapid, and the flue gases will reach the ground relatively near to the base of the chimney. When turbulence is weak, the gases drift away from the chimney in a thin compact plume, so that the point of maximum concentration at ground level is much farther downwind.

The effect of raising the height of a chimney is to reduce appreciably the concentrations of pollution at places within about ten chimney lengths of the base of the chimney, while increasing only very slightly the concentrations at large distances. Tests have been made over 30 years with chimney stacks of various heights at a number of American smelting and refining plants. They show that within 3 or 5 kilometres of the stack concentrations of sulphur dioxide may be reduced by from 25 to 50 per cent by doubling the height of the stack. At Selby, California, a smelter stack of 44·5 m was replaced by one of 184·5 m. The average concentration of sulphur dioxide at an observation post 4·2 km away was reduced to less than half at all seasons of the year. If there had been no pollution from other sources the reduction would probably have been more than two-thirds. None of the tests were sensitive enough to determine whether there was an increase in pollution at great distances when the stack height was raised.

In Great Britain there was a move towards a specification of the way in which the height of any particular chimney under design should be calculated. The Beaver Report in Appendix VI gave general advice, without formulating any precise rules. A working party of the Federation of British Industries considered the problem of small installations and its Chairman Mr. G. Nonhebel also considered larger installations. The joint suggestions may be summarized:

1. The calculated maximum concentration at ground level due to the chimney should not exceed 1·2 mg/m³ (0·4 ppm) averaged over 3 min, or 0·6 mg/m³ averaged over 3 hr.

2. In consequence of (1) boiler plants of aggregate capacity 2000–15,000 kg h^{-1} require chimneys ranging from 18 to 36 m high.

3. Chimneys for offices and other commercial buildings should extend at least 3 m above the ridge of the building, and the velocity of discharge of the flue gases should be high. Model tests in wind tunnels are recommended to help in choosing the position of the chimney.

4. The chimneys of boiler plants of aggregate capacity over 15,000 kg h^{-1} should be at least $2\frac{1}{2}$ times the height of surrounding buildings, including the building served by the chimney if this presents a long face to the wind, and should also satisfy condition (1).

These rules have been formulated as a nomogram and issued as Clean Air Act – Memorandum on Chimney Heights (revised 1967). The nomogram is a means of rapidly calculating H from the formula

$$H^2 = 9E/20P$$

where P is the maximum allowable concentration. This uncorrected chimney height is modified to take into account the character of the area (five classes range from rural to heavily industrial) and the size of nearby buildings. This nomogram is applicable to all emissions of sulphur dioxide between 1·36 and 816·0 kg h^{-1} and is at present the guideline used to determine required chimney heights in applications for planning permission in the U.K. The result is only strictly applicable to planar situations and cannot therefore be extended to valley sites nor indeed to "unusual" meteorological conditions, e.g. strong inversions, highly turbulent wind flows. "Experience" is thus usually added. Theoretically advances have been made in prediction of the behaviour of stack effluents in such situations but the conclusions are, as yet, not perfect.

CHAPTER 16

Air-Pollution Control—Law and Administration

IN MOST countries where formal legislative controls exist these comprise some form of planning consent or landuse strategy and more specific technical controls. This chapter will consider in detail the system operative within the United Kingdom as representative of a long-established system based on principles which equate both economic and technological feasibility, the concept of "best practicable means", with the increasing use of fixed emission standards of the European Economic Community and the United States of America. A detailed discussion of legislative and administrative provisions on a worldwide basis is not considered appropriate in any detail in a single chapter since not only has the past decade produced voluminous provisions but constant revision would lead to a need for rapid amendment to the text. It is suggested that the serious student will be able to use the information as a background to compare legislation not within the scope of this text.

The United Kingdom System

It seems appropriate to commence a detailed study of the United Kingdom system by stating its achievements as seen by the Royal Commission on Environmental Protection. In their fifth report the Commission found the present system achieved a great deal, emission of smoke from industrial processes in the U.K. went down by a remarkable 96 per cent over the period 1956–1973. Smoke from domestic fires accounted for 94 per cent of all smoke emitted but where smoke control has been introduced the improvement in air quality has been dramatic and although sulphur dioxide emission rose slightly during the same period ground-level concentrations fell by 45 per cent in urban areas, although in some rural areas ground level concentrations rose.

The United Kingdom system can be said to have developed over some 600 years being at times somewhat sporadic and local in nature and with little concerted effort. The first real attempt to control smoke came in 1875 with the Public Health Act of that year which introduced the concept of smoke abatement whilst as early as 1863 the Alkali Act of that year had recognized the special problems associated with certain industrial non-combustion processes and introduced the now accepted concept of "best practicable means" and the Alkali Inspectorate who were to administer its provisions, the remit of which has gradually increased with the passage of the current Alkali, etc., Works Registration Act, 1906.

It was in this form that subsequent and current legislation was to be formulated with the concept of a dual inspectorate comprising local government and central government branches, the latter in the words of the Beaver Committee Report, to be responsible in the case of certain industrial processes in which the prevention of dark smoke, grit or harmful gases present special technical difficulties, an objective redefined by the Royal Commission on Environmental Pollution in 1976.

The Alkali Inspectorate—structure and responsibilities

Since the passing of the Control of Pollution Act, 1974 there has been set up a central government body in the form of the Health and Safety Executive embracing the Alkali and Clean Air Inspectorate together with other central government inspectorates.

The Alkali and Clean Air Inspectorates with the Health and Safety Executive administer the Alkali Acts as amended by the Control of Pollution Act, 1974 and the Health and Safety at Work Act, 1974.

The Alkali Acts essentially control some sixty-one scheduled processes together with certain noxious or offensive gases which are associated with them. In controlling these processes a legislative philosophy has evolved which may be summarized as:

1. Annual Registration of the processes.
2. Prior to initial registration the inspectorate must be satisfied that the best practicable means for preventing the escape of noxious or offensive gases or rendering them harmless or inoffensive prior to discharge.
3. Maintenance of the best practicable means.
4. Where quality standards are set for effluent gases, i.e. total acidity, ensuring compliance.

The actual operation of control is based on the operation of the concept of best practicable means within the framework in 1–4 above inasmuch as the owner *must* (Alkali Act, 1906, Section 7 (1), Health and Safety at Work Act, 1974, Section 5) use the best practicable means for preventing the escape of noxious or offensive gases, for preventing their discharge to the atmosphere and for rendering gases where discharged harmless and inoffensive. It is not sufficient to operate on the basis of discharge to atmosphere at a sufficient height to be harmless and inoffensive unless each parameter is met, i.e. only after all other reasonably practical methods have been used can consideration be given to emission of gases.

The use of this approach has led to the necessity to define what can be considered reasonably practicable and what if any shall be the quality of any emissions which will inevitably arise from selected processes.

In the context of both the Alkali and Clean Air Acts, reasonable has been held to mean "reasonably practicable having regard among other things to local conditions and circumstances, to the financial implications and to the current state of technical knowledge" (Clean Air Act, 1956, Section 34).

It is within this framework that the Alkali and Clean Air Inspectorate supervise works falling within their jurisdiction and have over a period of time introduced presumptive limits of emission in addition to those defined in statute controlling acidic emission. These so-called presumptive limits, so called because compliance presumes that the best practical means of control is in operation, are less rigid in application and take into consideration such matters as technological change and evidence from other bodies whether medical or otherwise on what are acceptable maxima for exposure (in occupation terms referred to as a Threshold Limit Value). The second central government inspectorate having an involvement to a lesser, but not insignificant, degree is the factory inspectorate. Their function is more broadly based and is concerned with the factory working environment and the control of dusts, gases and toxic fume falls within their scope. This aspect of control will doubtless increase as a consequence of the Health and Safety at Work Act, 1974 and the reader's attention is directed to the voluminous regulations made currently under the provisions of the Factories Act, 1961 which space dictates cannot be adequately reproduced in a volume of this size.

In concluding the considerations of central government control it can be seen that the provisions are based on environmental impact versus what can be achieved in terms of economic and engineering terms and to an extent this philosophy is the unifying theme of the Clean Air Acts administered by local government.

Local Government Administration and Control

Local government control falls within the jurisdiction of the broadly based Environmental Health Inspectorate who are responsible to their local government employers for the enforcement of the Clean Air Acts and the relevant provisions of the Public Health Acts, Control of Pollution Act, 1974 and the Health and Safety at Work Act, 1974.

A detailed section by section consideration of all this legislation would similarly not be possible in a volume dealing with both the technological, meteorological and administrative provisions but the authors propose to consider such provisions as will illustrate the legislative philosophy.

The Clean Air Acts

The main objective of the Acts are:

1. The control of smoke emissions from any chimney or industrial and trade premises other than from a chimney subject to certain exemptions.
2. The control of new furnace installations by compulsory notification to the local authority with a system of prior approval where necessary.
3. The control of unacceptable ground-level concentrations of effluent gases, particularly SO_2, by a system of approval of chimney height.
4. The control of emissions of particulate matter from chimneys.
5. The control of low-level smoke emissions from domestic sources by empowering local authorities to make smoke-control areas.

6. The Acts of 1956 and 1968 are both considered to be enabling acts since their provisions make possible the bringing into effect regulations to deal with detailed technical matters, standards and exceptions.

Smoke Emissions

Section 1 of the 1956 Act makes it an offence to allow dark smoke to be emitted from the chimney of any building whilst sections 19 and 20 impose similar restrictions on railway engines and ships which are in waters not navigable by sea-going ships and in certain waters within the seaward limits of the territorial waters of the United Kingdom.

The Act defines dark smoke with reference to the Ringelmann Chart as being so dark as or darker than Shade 2 (Clean Air Act, 1956, Section 34) and this definition is used to facilitate the enforcement of regulations made under the Act known as the Dark Smoke (Permitted Periods) Regulations, 1958 and the Dark Smoke (Permitted Periods) (Vessels) Regulations of the same year. The effect of these regulations are to impose limits on the time which dark smoke may be emitted for the carrying out of the operation known as soot-blowing and having regard to the number of furnaces discharged into the flue whilst the regulations concerned with vessels impose similar periodic restraints on ships with various boiler plant and undergoing various operations. In both cases specific provisions are made to prohibit the emission of black smoke (smoke as dark as or darker than Shade 4 when compared to a Ringelmann Chart) for much shorter periods of time. The reader is directed to the regulations for details of the time periods which may well undergo revision as a result of increased fuel combustion efficiency, although the authors are not aware of any proposed imminent change.

In addition to the concessions granted by regulations the Act (Clean Air Act, 1956, Section 1 (3)) permits certain defences which may be accepted in proceedings for an offence. The reader will appreciate that these defences are based on basic principles of combustion and if the reader is in doubt as to the relevance of these provisions a revision of earlier material is essential.

The defences are:

(a) lighting up a furnace from cold and taking care to prevent or minimize the emission of smoke;
(b) failure of the apparatus not reasonably foreseen;
(c) lack of suitable fuel provided that the least unsuitable fuel which could be obtained was being used;
(d) a combination of the above.

The control did not extend to the emission of smoke other than from a chimney and any control of such emissions had to be pursued through Public Health Act legislation and its statutory nuisance provisions. This situation was amended by the Clean Air Act of 1968 whereby, subject to prescribed exemptions, the emission of dark smoke from industrial or trade premises, other than from a chimney, was prohibited (Clean Air Act, 1968, Section 1). The control was mainly to prevent excessive smoke emission caused by the indiscriminate burning of industrial or trade refuse in the open, but the section is also a control on the emission of dark smoke from industrial processes not having a chimney. Exemption regulations have been made by the Secretary of State in the form of the Clean Air (Emission of Dark Smoke) (Exemption) Regulations, 1969 which exempt the emission of dark smoke caused by the burning of the following materials or classes of container:

(a) Timber and waste material from demolition except synthetic rubber, flock or feathers.
(b) Explosive waste.
(c) Matter burnt as a consequence of research or training in fire control.
(d) As a consequence of road surfacing or similar.
(e) Animal carcasses slaughtered or having died as a result of disease.
(f) Pesticide wastes and containers.

Attached to these classes of material are certain conditions basically requiring that there is no other reasonably safe and practicable method, emission of dark smoke is minimized and supervision is ensured, the conditions being applied where they are relevant and practicable restraints to the process under consideration.

To enable local control of smoke-emission, provisions have been made in the 1956 Act for all new furnaces to be as far as possible smokeless in operation. All new installations must be reported to the local authority with the exception of certain low-rated appliances and within the framework of this provision is a requirement that if so required a local authority must after the examination of plans and a specification, if satisfied, confirm that if correctly operated and maintained the furnace is capable of smokeless operation. To enable control of bonfires, etc., not on industrial premises provisions of the Act (Section 16, Clean Air Act, 1956) extends statutory action to smoke sufficient to cause nuisance to the neighbourhood from sources otherwise uncontrolled.

A final legislative provision to assist in the control of smoke emission is that of power to make regulations concerning smoke-density meters and recorders with requirement to make any recordings available to the local authority. No regulations have yet been made by the Minister.

The Control of Chimney Heights

The philosophy behind the control of chimney heights is that at whatever height smoke and flue gases are discharged gravity will eventually bring the larger practicles of grit, dust and soot to the ground and because of atmospheric turbulence some of the smaller suspended particles will reach the ground. The higher the point of discharge the more dilute will be the effluent gases and particles by the time they reach ground level where it is proposed concentrations will not become dangerous.

This philosophy has been translated into legislation which permits a local authority to require a minimum height of discharge of a chimney after a consideration of such matters as would lead to a higher ground-level concentration of the representative index sulphur dioxide. The essential processes of control are;

(a) from a knowledge of fuel type, sulphur content and rate of fuel consumption an uncorrected height is obtained related to mass rate of sulphur dioxide emission;
(b) the uncorrected height is related to the geographical location of the installation i.e. proximity of other polluting sources and the geometry of the chimney and attached buildings either of which may give rise to downdraught. The result of this comparison gives rise to a final or corrected chimney height.

The principal legislative controls concerned with the point of discharge of combustion gases are:

(a) Clean Air Act, 1956, Section 10 controlling chimney heights in respect of non-combustion processes.
(b) Clean Air Act, 1968, Section 6 applies to chimney height approval to processes serving combustion processes, within certain combustion ratings.
(c) The Building Regulations, 1976 which impose requirements on the positioning of flue outlets.
(d) The Public Health Act, 1961 which makes provision when construction of a new building overreaches an adjoining chimney.

In addition to the reduction of ground level concentrations by chimney height control the provisions of the Control of Pollution Act, 1974 has resulted in the making of regulations to limit the sulphur content of oil fuels, in parallel with many other member states of the European Economic Community.

Grit and Dust from Furnaces

The underlying principle in relation to particulate emissions is that of best practicable means and as a reflection of the policy of requiring modifications within the limitations of economies of scale-arrestment plant was only necessary in situations of major potential emission, e.g. where pulverized fuel or solid fuel at a high rate was burned. Subsequent control has been extended by the Clean Air Act, 1968 and by regulations (The Clean Air (Emission of Grit and Dust from Furnaces) Regulations, 1971) to control emissions from plant burning both pulverized fuel and solid fuel at a lower rate of 100 lb/hr (45.36 kg h^{-1}) or more or liquid fuel at a rate of $1 \cdot 25 \times 10^6$ B.t.u. an hour (1.185×10^6 kJ h^{-1}) any liquid or gaseous matter.

Regulations specify the amount of grit permitted in the effluent gases which are to be read together with the provisions relating to the requirement to install and maintain grit and dust arrestment plant, the latter being the subject of prior approval provisions.

Exemptions from the necessity to provide arrestment plant are provided both by Regulations in the form of the Clean Air (Arrestment Plant) (Exemption) Regulations, 1969 and direct exemption by the local authority in respect of a specific furnace. The principal exemptions relate to temporary installations.

Provisions have been made by regulation as a consequence of both Clean Air Acts which prescribe the procedure to be followed by a local authority requiring an industrialist to make grit and dust measurements. The provisions of these Clean Air (Measurement of Grit and Dust) Regulations, 1971, set out in detail the technical and administrative provisions but it should be noted that measurements are both costly and time consuming and local authorities are unlikely to entertain indiscriminate action.

The reader's attention is particularly directed to the control of grit and dust emission as an excellent example of the best practicable means approach where costs are high and unless the correct technological applications are used results may be less than satisfactory.

Smoke-control Areas

The control of low-level smoke emissions was originally in the hands of a local authority having to promote its own special act or for the inclusion of special provisions in a local act to set up zones (smokeless zones) in which emission of smoke was prohibited. With the passing of the Clean Air Act, 1956 and subsequent provisions of the Clean Air Act, 1968 and Housing Act, 1964 the concept of smoke-control areas was given legislative form.

The principal provisions of the legislation is to permit local authorities to make a Smoke-control Order which must be submitted to the Secretary of State for the Environment for confirmation. When operative it is an offence for an occupier of premises to allow smoke emission from a chimney unless the smoke is caused by the use of an "authorized fuel" (a fuel which is inherently low in volatile content or has had a high proportion of the volatile material removed to enable smokeless combustion), or the fireplace which the chimney serves is exempt from the Order. Authorized fuels are designated by regulation as are classes of fireplace and the reader is advised to consult the most current orders since these are constantly updated.

The 1968 Act extends the powers of the Minister by enabling him to require a local authority to make a Smoke-control Order and by making it an offence to sell for delivery to premises within a Smoke-control Area any other than an authorized fuel, other than for exempt fireplaces.

The effect of the area is to enable the local authority to pay the owners or occupiers of a dwelling house a grant to enable the cost of fireplace adaptations to be undertaken, the eligible expenses being set by central government and issued to the local authorities by circulars.

The payment of grant is subject to criteria set by the local authority as to the satisfactory nature of the work and that the appliance is not designated (designated by the provisions of Section 95 of the Housing Act, 1964) as an appliance for which suitable fuels are not fully available.

As has been previously stated where smoke control has been introduced the improvement in air quality has been dramatic (Royal Commission on Environmental Pollution 5th Report, 1976).

In furtherance of controls which the United Kingdom provisions of the Control of Pollution Act, 1974 and regulations made thereunder permit local government Authorities to commit expenditure to research into air pollution, require occupiers of premises to make returns on emissions, arrange for the publications of information.

AIR-POLLUTION LEGISLATION THROUGHOUT THE WORLD

A chapter devoted to legislation would be incomplete without a reference to the legislative forms of other countries but such is the volume of legislation that this text cannot hope to cover any but the essentials and, as previously stated, only the European Economic Community and the United States will be discussed to provide the reader with a background for comparison.

Air Pollution Control Within the Member States of the European Economic Community

It should be appreciated that the Treaty of Rome makes no detailed reference to the environment although the Heads of State at a meeting in 1972 agreed on a Programme of Action on the Environment.

Following its adoption in 1973 a research programme was drawn up to determine dose-effect relationships of pollutants on both man and other significant targets. On the results the Commission has and will continue to draw up environmental quality standards which may result in an increase in the imposition of uniform emission standards within the member states.

The current situations within the member states must therefore be viewed in the light of a continuing input from the Commission.

Belgium

The system of control approximates the United Kingdom system where industrial provisions and domestic combustion processes are largely separated with the greater attention being paid to the larger industrial appliances.

In a similar method to that adopted in the United Kingdom's system of scheduled processes industrial installations are classified according to their polluting potential and legislative controls specify chimney height, arrestment and cleaning plant and general process hygiene to be observed in their operation. To facilitate control of high sulphur dioxide build up prohibitions exist for the burning of high volatile fuels or fuels with a sulphur content of more than 1 per cent, similar prohibitions relate to the indiscriminate burning of waste material.

At the domestic level control is effected by requiring all appliances to be capable of satisfactory operation unless the appliance is situated in an area of special protection where existing sulphur dioxide levels present an existing problem, in these areas more specific controls are applicable.

The Belgian system is increasingly moving towards a system of fixed emission standards accompanied by considerably more monitoring by enforcement agencies as opposed to the convention of the discharger being required to monitor.

Denmark

Controls rely on joint application of land-use planning together with environmental protection measures of a more specific nature.

The legislation dealing with planning is concerned with the control of industrial development and ensuring siting of the potential polluting processes in areas where proximity to similar sources or topography would significantly increase ground-level concentrations. The planning process operates in conjunction with environmental protection provisions relating to polluting processes whereby prior approval of plant is necessary and enforcement provisions relating to remedial works on existing plant. The provisions mainly relate to controls on the pre-treatment of effluent gases and hence regulation of the quality of the discharge. At the time of writing the Minister of the Environment has made specific regulations controlling sulphur content of fuels and legislation to control chimney height.

France

The system currently in operation relies on joint measures controlling land-use planning by siting establishments so as to prevent excessive pollution problems and in certain areas of existing pollution and high density of population, designated "zones of special protection", qualitative standards are imposed.

Essentially areas for development are classified as:

(i) first class where proximity to residential accommodation is the determining factor;
(ii) second class where operation is dependent on specified control measures;
(iii) third class where operation is permitted subject to restraints made to protect public health.

Control of appliances rather than the operation relies on orders prescribing technical specifications commensurate with satisfactory performance whilst chimney-height control operates as a primary means of reduction of ground level build up of pollutants.

Enforcement is via the Police Judiciaire and penalties on conviction are by imposition of fines or imprisonment.

Federal Republic of Germany

The system closely follows that of the United Kingdom in that an enabling law, the Federal Commission Control Law, operative since 1974, gives power to make appropriate regulations. The power to make regulations has not been invoked to any great extent but it has provided for a system of licensing of certain plants and imposes design criteria for plants likely to give rise to emissions. Where licences are required and conditions are attached specifying emission standards two types of discharge limits are recognized, namely "nuisance limits" and "emission limits"; the former relating to emissions in a particular situation having regard to effect on living systems whereas the latter are based on what is achievable within the limits of technology, a system not unlike that of the best practicable means approach of the United Kingdom.

In addition to these provisions legislation exists for control of chimney construction and making provision for measurement of emissions by the industrialist. It should be noted that these specific provisions are used in conjunction with planning controls designate to segregate incompatible developments.

Ireland

Whilst there are legislative differences it is not surprising that the administrative system and controls are similar to those operated in the United Kingdom.

Essentially control is by means of land-use planning and both legislative and licensing controls. The provisions of the Alkali, etc., Works Registration Act, 1906 applies but without subsequent amendments operative in the United Kingdom.

Italy

The system of control follows the general pattern using land-use planning as a criterion for siting of potential polluters by taking into consideration the topographical, meteorological and population parameters. The system operates in conjunction with specific limitations on plant operation and fuel types with controls being enforced where volatile and sulphur content exceed specified limits. Discharges are limited by Regulation and in terms of industrial plant, as opposed to purely heating plant, maximum permissible ground-level concentrations are imposed with the onus of monitoring being placed on the operator although powers exist for public authority monitoring and intervention.

Luxembourg

Luxembourg is currently poorly provided with legislative controls although current proposals are for the passage of a Bill establishing legislation to prevent air pollution.

The Netherlands

The Netherlands adopt a policy not unlike the other member states with both planning and clean air legislation. The Physical Planning Act provides for a planning strategy whereby industrial land-use zoning is implemented, whereas the Air Pollution Act, 1970 imposes a somewhat unusual system of financial levy on pollutant discharge determined on a quantitative and qualitative fuel use basis. Where maximum concentrations are set these are determined having regard to best practicable means and these are used in conjunction with a system of licensing to gain control over missions.

Since 1974 legislation has been in force, with progressive amendments, limiting the amount of sulphur permitted in fuel types.

The United States of America

Control of Air Pollution in the United States dates from about 1881 when Chicago and Cincinnati led the way with smoke-control laws in 1881. By 1912 twenty-three out of the twenty-eight cities with populations greater than 200,000 had similar laws although specific State enabling legislation was sometimes needed and few States involved themselves with control programmes, this situation remained until the mid-fifties.

Federal involvement became significant in the early fifties when it became apparent that automotive emissions constituted the major constituents of Los Angeles, problems. This emphasis on the involvement of gaseous pollutants and a wider appreciation of the problem not being purely local involved all three levels of government and the enactment of the Federal Clean Air Act, 1963 which has subsequently been amended, with major amendments in 1970 and 1977.

The major controls set up under the Clean Air Act, 1963 was the imposition of air-quality standards to be determined by the Environmental Protection Agency. In the first instance air-quality standards for six of the most prevalent air pollutants were set. The standards related to particulate matter, sulphur oxides, carbon monoxide, hydrocarbons, nitrogen dioxide and photochemical oxidants which were known as the "criteria" pollutants. The quality standards are divided between primary and secondary types, the former based on criteria which will allow an adequate margin of safety to protect the public health whilst the latter are based on such criteria as is requisite to protect the public welfare from known or anticipated adverse effects associated with the presence of such air pollutants in the ambient air. The Act requires that every State shall have the responsibility for ensuring compliance, within their total area, of any Standards and to facilitate this every State must prepare and submit to the central administration an implementation plan setting out a strategy to deal with both primary and secondary targets. In return the Act places a duty on the Federal government to establish and raise the standards and in so doing the air-quality criteria shall "accurately reflect the latest scientific knowledge useful in identifying effects on public health or welfare" (note the resemblance to best practicable means). Simultaneous with the issuance of criteria there is also a requirement that the Administrator issue information on technology and costs of emission control including data on alternative fuels, processes and operating methods. There are time limits for state authorities to implement a control plan subject to permitted extensions, and it should be noted that a ground for extension is that a source or class is unable to comply with any requirement because the necessary technology or other alternative methods of control are not available. Where no such "exemption" is in force it is an offence for any owner or operator of any new source to operate it in violation of any standard and also it is an offence to modify any existing source which will emit an air pollutant to which any standard applies unless the source if properly operated will not cause an illegal emission. In the case of an existing source the standard will not apply for a 90-day period after implementation but this period may be extended to 2 years provided no imminent danger to health is expected.

Enforcement of these standards is by Federal or State action who may by order require the compliance with any particular course of action or modification, non-compliance with which can result in a daily fine or imprisonment.

The deadline for achieving public health (primary) standards for ambient air quality was set for mid-1975 (Clean Air Amendments, 1970) but by 1975 the levels had only been fully achieved in 91 out of 247 Air Quality Control Regions but it was noted that of the 20,000 largest stationary sources, accounting for 85 per cent of all stationary source pollution, 15,600 were in compliance with emission regulations.

The programme of compliance with standards continues within the United States and the interested reader would be advised to consult the United States annual report of the Council on Environmental Quality.

Motor Vehicle Pollution

It is now generally accepted that motor vehicles are an important source of atmospheric pollution especially in urban areas. It is for this reason that whilst space prevents an exhaustive approach, the United Kingdom Controls, as amended by European legislation, will be discussed with mention of the American control measures.

Control Action has been taken on a tripartite form:

(a) motor fuel,
(b) vehicle construction standards,
(c) vehicle in use standards.

Regulations related to motor fuel composition are made pursuant to the Control of Pollution Act, 1974 with lead in petrol being of primary concern to the extent that the Motor Fuel (Lead Content of Petrol) Regulations have been made to ensure a lowering of lead content in the fuel, this being in accord with the requirements of the EEC Directive (78/611/EEC) which prescribes a maximum lead content. In the context of diesel fuel sulphur is the predominant consideration and regulations in the form of The Motor Fuel (Sulphur Content of Gas Oil) Regulations, 1976 specify a progressive reduction in sulphur content in the period 1976–1980.

In addition to fuel controls vehicle-construction standards are imposed under the provisions of the Road Traffic Act, 1972 and these are principally concerned with the regulation of particulate and gaseous emissions which are now the subject of EEC Directives to control gaseous emissions and recently to extend control to include crankcase emissions, carbon monoxide emissions at idle, nitrogen oxides and carburation controls. The major controls in respect of diesel vehicles have been the control of smoke emissions and these controls have been reinforced by European standards for smoke opacity.

The situation in the United States of America for control of mobile sources is similar in respect of fuel controls and certain fixed emission standards are applicable to classes of new motor vehicles or new motor vehicle engines. These provisions are made under the Federal Clean Air Act, 1963. The American system of control designates target dates for the reduction of emissions of hydrocarbons, carbon monoxide and nitrogen oxides which have been extended since a shift from control of gaseous emissions has been sacrificed for research on automobile efficiency. In addition to emission standards transport control strategies form a significant future proposal in the American context, i.e. controls on parking and preconstruction controls of potential large sites of vehicle concentration.

The current controls of motor vehicle emissions are principally constructional in nature augmented with fuel restrictions but it is suggested that future controls may well rely on land-use planning and motor vehicle access restrictions. This area of pollution control is rapidly expanding and the reader is directed to the constant review of emission standards.

Appendix A
Conversions

Concentrations

	ppm	μg m^{-3}	at 0°C, 10^5 N m^{-2}
SO$_2$	1	2860	
CO	1	1250	
O$_3$	1	2141	
PAN	1	5398	
			at 25°C, 10^5 N m^{-2}
SO$_2$	1	2620	
CO	1	1145	
NO	1	1230	
NO$_2$	1	1880	
O$_3$	1	1962	
PAN	1	4945	

Length

1 m = 3·2808 ft = 39·3701 in.

Volume

1 ft^3 = 0·0283 m^3
1 gallon (U.K.) = 4·546 l
1 gallon (U.S.) = 3·785 l

Mass

1 kg = 2·204 62 lbm

Force and pressure

1 N = 1 kg m s^{-2}
4·448 22 N = 1 lbf
1 lbf in.$^{-2}$ = 6894·76 N m^{-2}

Appendix A. Conversions

Energy and power

1 B.t.u. = 252·0 cal = 1055·1 J
1 cal = 4·1868 J
1 B.t.u. h^{-1} = 1·055 kJ h^{-1} = 0·293 W
1 J kg^{-1} = 4·299×10^{-4} B.t.u. lb^{-1}
1 B.t.u. ft^{-3} = 37·263 kJ m^{-3}
1 B.t.u. lb^{-1} = 2·326 kJ kg^{-1}
1 therm = 10^5 B.t.u.
1 mtoe = 4·25×10^{13} B.t.u. = 4·48×10^{16} J
1 mtce = 2·55×10^{13} B.t.u. = 2·69×10^{16} J,
 i.e. 1 tonne oil = 1.7 tonnes coal = 425 therms
1 kWh = 3·6×10^6 J = 3416 B.t.u.
1 h.p. = 746 W
1 tonne coal equivalent = 7·32×10^3 kWh = 250 therms

Appendix B
British Standards

Pollution and Measurement

B.S. number

893:	1940	Method of testing dust-extraction plant and the emission of solids from chimneys of electric power stations.
1747:		Methods for the measurement of air pollution
	Part (1): 1969	Deposit gauges.
	Part (2): 1969	Determination of concentration of suspended matter.
	Part (3): 1969	Determination of sulphur dioxide.
	Part (4): 1969	The lead dioxide method.
	Part (5): 1972	Directional dust gauges.
1756:	1952	Code for the sampling and analysis of flue gases: Indicators and recorders.
2740:	1956	Simple smoke alarms and alarm metering devices.
2741:	1957	Recommendations for the construction of simple smoke viewers.
2742C:	1958	Ringelmann chart.
2742M:	1960	Miniature smoke chart.
2811:	1969	Smoke density indicators and recorders.
3048:	1958	Code for the continuous sampling and automatic analysis of flue gases: Indicators and recorders.
3250:		Methods for the thermal testing of domestic solid fuel burning appliances with convection
	Part (1): 1960	Flue loss method.
	Part (2): 1961	Hood method.
3405:	1971	Simplified methods for measurement of grit and dust emissions from chimneys.
3841:	1972	Measurement of smoke from manufactured solid fuels for domestic open fires.

Bibliography

ANNUAL CONFERENCES OF THE NATIONAL SOCIETY FOR CLEAN AIR.
Annual Reports on Alkali, etc., Works, HMSO, London.
BALL, D. J. and BERNARD, R. E. Evidence of photochemical haze in the atmosphere of Greater London. *Nature*, **271**, 733–734, 1978.
BARNES, R. A. and EGGLETON, A. E. J. The transport of atmospheric pollutants across the North Sea and English Channel. *Atm. Environ.* **11**, 879–892, 1977.
BEAVER COMMITTEE ON AIR POLLUTION. *Interim Report*. December 1953, and *Report*, November 1954, HMSO, London.
BENNETT, R. J., CAMPBELL, W. J. P. and MAUGHAN, R. A. **Changes in Atmospheric Pollution Concentration in Mathematical Models for Environment Problem,** ed. C. A. Brebbia, Pentech Press, 1976.
(The) Boiler Operator's Handbook, NIFES, 1969.
BOSANQUET, C. H. and PEARSON, J. L. The spread of smoke and gases from chimneys. *Trans. Faraday Soc.* **32**, 1249–1264, 1936.
BOUHUYS, A, BECK, G. J. and SCHOENBERG, J. B. Do present levels of air pollution outdoors affect respiratory health? *Nature*, **276**, 466–471, 1978.
Brickworks Valley; a study of pollution (not including fluorine) in report for Bedfordshire County Council Health Department. Nat. Soc. Clean Air, London, *Smokeless Air*, Winter 1961
BRIGGS, G. A. *Plume Rise*, AEC Critical Review Series, U.S. Atomic Energy Commission, 1969.
BRIMBLECOMBE, P. Air pollution in industrializing England. *JAPCA*, **28**, 115–118, 1978.
BURROWS, B., KELLOGG, A. L. and BUSKEY, J. Relationship of symptoms of chronic bronchitis and emphysema to weather and air pollution. *Arch. Environ. Hlth.* **16**, 406–413, 1968.
BUTLER, J. D. and MACMURDO, S. D. Interior and exterior atmospheric lead concentrations of a house situated near an urban motorway. *Int. J. Environ. Studies*, **6**, 181–184, 1974.
CAIN, W. S. To know the nose: keys to odour identification. *Science*, **203**, (2), 467–469, 1979.
CAMPBELL, I. M. *Energy and the Atmosphere, A Physical-Chemical Approach*, John Wiley & Sons, 1977.
CAPLAN, K. J. All about cyclones. *Air Engineering*, Sept. 1964.
CARROLL, J. D., CRAXFORD, S. R., NEWALL, H. E. and WEATHERLEY, M-L. P. M. Trends in the pollution of the air of Great Britain by smoke and sulphur dioxide 1952–59. *Proceedings of the Clean Air Conference*, Harrogate, 1960, published by the National Society for Clean Air.
CHANDLER, T. J. *The Climate of London*, Hutchinson, 1965.
CHANDLER, T. J. and ELSAM, D. M. Meteorological controls upon ground level concentrations of smoke and sulphur dioxide in two urban areas of the U.K. *Warran Spring Laboratory* LR 245 (Ap).
CLARKE, A. G., GASCOIGNE, M., HENDERSON-SELLERS, A. and WILLIAMS, A. Modelling air pollution in Leeds (U.K.), *Int. J. Environ Studies*, **12**, 121–132, 1978.
Clean Air Act, 1956 HMSO, London.
Clean Air Act, 1968, HMSO, London.
CLEGG, L. and SHORT, W. Grit and dust emissions and means for control. *Proc. Nat. Soc. Clean Air Conf.*, 1963.
CRAXFORD, S. R. Air pollution, past, present and future—a study in trends. *Inst. Petrol. Rev.*, May 1961.
CRAXFORD, S. R. and BAILEY, D.L. R. The acidity of rain and the sulphur dioxide in the air in country districts in the U.K. and Ireland. *Report, Warren Spring Laboratory*, HMSO, London, 1970.
CRAXFORD, S. R., SLIMMING, D. W. and WILKINS, E. T. **The Measurement of Atmospheric Pollution: the accuracy of the instruments and the significance of the results,** *Nat. Soc. Clean Air*, 1960. Clean Air Conference.
CRAXFORD, S. R., WEATHERLEY, M-L. P. M. and GOORIAK, B. D. **National Survey of Air pollution 1961–71.** Vol. 1. Part 2. *The United Kingdom, a Summary*, HMSO, London, 1972.
CSANADY, G. T., *Turbulent Diffusion in the Environment*, Reidel, 1973.

DAGNE, R. R. *J. Wat. Pollut. Control Fed.* **44**, (4), 583–594, 1972.
DALHAMN, T. Effect of cigarette smoke on ciliary activity. *Amer. Rev. Resp. Dis.* **93**(3), 108–114, 1966.
DANARD, M. B. Numerical modelling of carbon monoxide concentration near highways. *JAM*, **11**, 947–957, 1972.
DEPARTMENT OF ENERGY STATISTICAL BULLETIN, *Energy Trends*, issued monthly.
Department of the Environment Odours Report of the Working Party on the Suppression of Odours from Offensive and Selected Other Trades, Part 1 (1974), Part 2 (1975), Warren Spring Laboratory.
DEPARTMENT OF THE ENVIRONMENT, *Digest of Environmental Pollution Statistics*, HMSO, London, 1978.
DIAMANT, R. M. E. *The Prevention of Pollution*, Pitman, 1974.
DRAVNICKS, A., PROKOP, W. H. and BOEHME, W. R. Measurement of ambient odours using dynamic farad-choice triangle olfactometer. *JAPCA*, **28**(11), 1124–1130, 1978.
(The) Efficient Use of Fuel, 2nd ed., HMSO, 1958.
ESSENHIGH, R. H. A short history of pulverized fuel firing. *J. Inst. Fuel*, Jan. 1961.
EVELYN, John. *Fumifugium* (reprinted by the National Society for Clean Air, 1960).
EWING, R. H. Potential relief from extreme urban air pollution. *JAM*, **11**, 1342–1345, 1972.
GARNER, J. F. and CROW, R. K. *Clean Air—Law and Practice*, 4th ed. Shaw & Sons Ltd., London, 1976.
GARNETT, A. Some climatological problems in urban geography with reference to air pollution. *Trans. Inst. Br. Geogr.* **42**, 21–43, 1967.
Gas Council Annual Report 1958/9.
GIBSON, J. The 1977 Robens Coal Science Lecture: The constitution of coal and its relevance to coal conversion processes. *J. Inst. Fuel*, 67–81, June 1978.
GIFFORD, F. A., Jr. Atmospheric dispersion. *Nuclear Safety*, **1** (3), 56, 1960.
GILPIN, A. *Control of Air Pollution*, Butterworth, 1963.
GREENE, R. Utilities scrub out SO_x. *Chem Eng.*, 101–103, 23 May, 1977.
HALES, J. M., WOLF, M. A. and DANA, M. T. A linear model for predicting the washout of pollutant gases from industrial plumes. *A.I.Ch.E. J.* **19**, 292–297, 1973.
HALSTEAD, C. A. Air pollution and relief in the Glasgow area. *Geoforum*, **19**, 67–72, 1973.
HAMILTON, P. M. Use of LIDAR in the study of chimney plumes. *Phil. Trans. Roy. Soc.* **265**, 153, 1969.
HARVEY, M. R. Natural gas firing of kilns-recent developments. *Claycraft*, **44**(12), 7, 1971.
HEALTH AND SAFETY EXECUTIVE, *H. M. Factory Inspectorate Threshold Limit Values 1976*, Tech. Data note 2/75, HMSO, London.
HECK, W. W., The use of plants as indicators of air pollution. *Int. J. Air Water Pollut.* **10**, 99, 1966.
HEIDORN, K. C. A chronology of important events in the history of air pollution meteorology to 1970. *Bull. Amer. Met. Soc.* 59, (12), 1589–1597, 1978.
HENDERSON-SELLERS, A. The relationship between air pollution and meteorology on Merseyside. Report submitted to County Planning Officer of Merseyside, May 1979.
HENDERSON-SELLERS, A. and SEAWARD, M. R. D. Monitoring lichen reinvasion of ameliorating environments. *Environ. Pollut.*, 19, 207–215, 1979.
HORSMAN, D. C., ROBERTS, T. M. and BRADSHAW, A. D. Evolution of sulphur dioxide tolerance in perennial rye grass. *Nature*, 276, 493–494, 1978.
INSTITUTE OF HEATING AND VENTILATION ENGINEERS (now Chartered Institute of Building Services) Guidebook C, *Fuels and Combustion.*
Investigation of Atmospheric Pollution, 31st Report, 1959.
KAISER, G. and FRYER, L. Modelling airborne radioactivity. *New Scientist*, 23 Nov. 1978.
LAWTHER, P. J. Air pollution and its effects on man. *The Sanitarian*, **69**, No.2.
LEACH, G., LEWIS, C., ROMIG, F., FOLEY, G. and VAN BAREN, A. A low energy strategy for the United Kingdom. *Science Rev. IIED*, London, 1979.
LEDBETTER, J. P. *Air Pollution*, Parts A, B, Marcel Dekker Inc., 1972.
LEONARDOS, G., KENDALL, D. and BARNARD, N. *JAPCA*, **19**(2), 91–95, 1969.
LUCAS, D. H. Application and evaluation of results of the Tilbury rise and dispersion experiment. *Atmos. Environ.* **1**, 421, 1967.
MCMULLAN, J. T., MORGAN, R. and MURRAY, R. B. *Energy Resources*, Edward Arnold, 1977.
MARTIN, A. E. Epidemiological studies of atmospheric pollution. *Bull. Min. Hlth, London*, **20**, Mar. 1961.
MARTIN, A. and BARBER, F. R. Investigations of sulphur dioxide pollution around a modern power station. *J. Inst. Fuel*, **39**, 294, 1966.
MEADE, A. *The New Modern Gasworks Practice*, Eyre & Spottiswoode, 1934.
MEDICAL RESEARCH COUNCIL, *Annual Reports*, HMSO, London.
MEETHAM, A. R. *Atmospheric Pollution in Leicester*, DSIR, HMSO, 1945, 1956.
MEETHAM, A. R. Natural removal of pollution from the atmosphere. *Q.J. Roy. Met. Soc.*, **76**, 359, 1950.

Bibliography

MEETHAM, A. R. Natural removal of atmospheric pollution during fog. *Q.J. Roy. Met. Soc.*, **80,** 96, 1954.
MEMORANDUM ON CHIMNEY HEIGHTS. *Ministry of Housing and Local Government*, HMSO, London, 1963.
Mortality and Morbidity during the London Fog of December 1952., Ministry of Health Report No. 95, HMSO, 1954.
MORTON, B. R., TAYLOR, G. I. and TURNER, J. S. Turbulent gravitational convection from maintained and instantaneous sources. *Proc. Roy. Soc.* A **234,** 1–23, 1956.
MURRAY, M. V. The shell boiler—an historical review. *J. Inst. Fuel*, No. 224, 1959.
NASRALLA, M. An investigation of some combustion generated pollutants. Ph. D. Thesis, University of Leeds, 1976.
NATIONAL SOCIETY FOR CLEAN AIR—*Annual Report.*
NATIONAL SOCIETY FOR CLEAN AIR, Brighton, England. History of air pollution in Great Britain. *Clean Air Year Books*, 1975 and 1976.
NATIONAL SURVEY OF AIR POLLUTION. *Annual Report.*
National Survey of Air Pollution, 1961–71, 2 parts.
NOLL, K. E. and MILLER, T. L. *Air Monitoring Survey Design*, Ann Arbor Science, 1977.
NONHEBEL, G. Recommendations on heights for new industrial chimneys. *J. Inst. Fuel*, **33,** 479, 1960.
NONHEBEL, G. Best practicable means and presumptive limits—British Definitions. *Atmos. Environ.* **9,** 709 –715, 1975.
OOMS, G. A new method for all calculations of the plume path of gases emitted by a stack. *Atmos. Environ.* **6,** 899–909, 1972.
OPEN UNIVERSITY PT272 Units 13–14. *Air Pollution.*
OPEN UNIVERSITY PT272 Unit 15. *Air Pollution Control.*
OECD, *Methods of Measuring Air Pollution*, Paris, 1964.
OECD Report on Programme on Long Range Transport of Air Pollutants, Paris, 1977.
OVERTON, J. H., ANEJA, V. P. and DURHAM, J. L. Production of sulphate in rain and raindrops in polluted atmospheres. *Atmos. Environ.* **13,** 355–367, 1979.
PARKER, A. *Clean Air J.* 4(14), 1974; *Clean Air Year Books*, 1975 and 1976, National Society for Clean Air.
PARKER, A. (ed.) *Industrial Air Pollution Handbook*, McGraw-Hill, 1978.
PARKER, H. W. *Air Pollution*, Prentice Hall, 1977.
PASQUILL, F. The estimation of the dispersion of windborne materials, *Met. Mag.* HMSO, **90,** 33, 1961.
PASQUILL, F. *Atmospheric Diffusion*, 2nd ed., Ellis Horwood Ltd., 1974.
PERKINS, H. C. *Air Pollution*, McGraw Hill, 1974.
PERRY, R. and YOUNG, R., eds. *Handbook of Air Pollution Analysis*, Halstead Press, 1978.
PINDARD, T. S. and WILKINS, E. T. *Annual Conference of the National Society for Clean Air*, 1958.
PLENDERLEITH, H. J. *The Preservation of Leather Bookbindings*, British Museum, 1946.
PYATT, F. B. An appraisal of tree sulphur content as a long term air pollution gauge. *Int. J. Environ. Studies*, **7,** 103–106, 1975.
Reducing Pollution from Selected Energy Transformation Sources, Graham & Trotman, 1976.
ROSE, H. E. and WOOD, A. J. *An Introduction to Electrostatic Precipitation in Theory and Practice*, Constable, London, 1966.
ROSS, R. D. *Air Pollution and Industry*, Van Nostrand Reinhold, 1972.
RUBEY, W. W. Geologic history of sea water; an attempt to state the problem. *Bull. Geol. Soc. Amer.* **67,** 1111–1148, 1951.
SCORER, R. S. *Air Pollution*, Pergamon Press.
SCORER, R. S. *Environmental Aerodynamics.* Ellis Horwood Ltd., 1978
SCORER, R. S. and BARRETT, C. F. Gaseous pollution from chimneys. *Int. J. Air Water Pollut.*, **5,** 1–15, 1961.
SCOTT, J. A. The London fog of December, 1962, *The Med. Officer*, **109,** 250–252, 1963.
SEARL, M. F. *Fossil Fuels in the Future*, U.S. Atomic Energy Commission, 1960.
SHEARS, W. J. C. Soot blowing and the Clean Air Act—some practical considerations. *J. Inst. Fuel*, No. 236, Sept. 1960.
SITTIG, M. *Environmental Sources and Emissions Handbook*, Noyes Data Co., 1975.
SKINNER, D. G. *The Fluidised Combustion of Coal*, M. & B. Monographs, Chemical Engineering (Mills & Boon Ltd., 1971).
SLAWSON, P. R. and CSANADY, G. T. On the mean path of buoyant bent over chimney plumes. *J. Fluid. Mech.* **28,** 311–327, 1967.
SLAWSON, P. R. and CSANADY, G. T. The effect of atmospheric conditions on plume rise. *J. Fluid Mech.* **47,** 33–49, 1971.

Statistical Year Book of the World Power Conference, No. 9, 1960.
STERN, A. C., ed. *Air Pollution* (3rd ed.), 5 vols., Academic Press, 1976.
STERN, A. C., WOHLERS, H. C., BOUBEL, R. W. and LOWRY, W. P. *Fundamentals of Air Pollution*, Academic Press, 1973.
SUTTON, O. G. A theory of eddy diffusion in the atmosphere *Proc. Roy. Soc. London*, A **135,** 143,. 1932.
SUTTON, O. G. The theoretical distribution of airborne pollution from factory chimneys. *Q. J. Roy. Met. Soc.* **73,** 426, 1947.
SUTTON, O. G. *Micrometeorology*, McGraw-Hill Book Co., New York, 1953, 333 pp.
TOUT, D. G. Manchester sunshine. *Weather*, **28,** 164–166, 1973.
TREVELYAN, G. M. *English Social History*, Longmans, Green, London, 1944.
TURNER, D. B. *Workbook of Atmospheric Dispersion Estimates*, U.S. Public Health Service Publication 999–AP–26, revised 1970 edition.
VICK, C. M. and BEVAN, R. Lichens and tar spot fungus *(Rhytisma acerinum)* as indicators of sulphur dioxide pollution on Merseyside. *Environ. Pollut.* **11,** 203–216, 1976.
Weather, April 1955.
WEATHERLEY, M-L, P. M. *Fuel Consumption, and Smoke and Sulphur Dioxide Emissions, in the United Kingdom up to 1976*, Warren Spring Laboratory LR 258 (AP), 1977.
WILKINS, E. T. Air pollution and the London fog of December 1952. *J. Roy. Sanit. Inst.* **74,** 1–21, 1954.
WILLIAMSON, S. J. *Fundamentals of Air Pollution*, Addison-Wesley, 1973.
WILSON, D. G. Alternative automobile engines. *Sci. Amer.* **239**(1), 27–37, 1978.
WOOD, C. and LEE, N. Cities and pollution. *Int. J. Environ. Studies*, **8,** 293–300, 1976.
WOOD, C. M., LEE, N., LAKER, J. A. and SAUNDERS, P. J. W. *The Geography of Pollution*, Manchester Univ. Press, 1974
World Energy Supplies 1970–1973, Statistical papers, Series J, No. 18, United Nations, New York, 1975.
20 Years of Air Pollution Control, Manchester Area Council for Clean Air and Noise Control.

Index

Acid
  rain 112
Aga cooker 106
Air
  composition of 108
  conditioning 201
  pollution (see Atmospheric pollution)
Aircraft 50
Alcohol 39
Alkali
  Acts 207
  Inspectorate 207
Animals 185, 188
Anthracite 19
Approved list of domestic appliances 98
Ash
  distribution in Britain 147
  emission 111
  measurement 130
  prevention 195
  spheres in grit 141
Atmosphere
  origin of 6
Atmospheric
  gases 108
  pollution 1
    and climate 113
    changes 164
    control 4
    cost 192
    deposit 163
    distribution 147
    effects 181
    emitted 109
    from cooling towers 50
    from engines 48
    from furnaces 88
    from kilns 94
    from paint particles 117
    from motor vehicles 218
    from soot blowing 79
    from spoilbanks 119
    history 3
    law 207
    measurement 123
    National Survey 160
    odours 115
    prevention 194
    radioactive 115
    rate of production 2
    statistics 162
    surveys 157
    suspended matter 110
    trends 160
    types 2
    variability 162
Automatic filter 128

Belgium 214
Best practicable means 207
Bitumen 18
Bituminous coal 18
Blast furnace 86
  checks and slips 87
  gas 41
Boghead 19
Boiler
  coal-fired 57
  deposits 78
  instruments 67
Boilers
  Cochran 58
  Economic 60
  hot water 64
  Lancashire 59
  shell 57
  thermal storage 61
  vertical 58
  water-tube 57, 62
Brick kilns 90
Briquettes 18, 38
British standards 222
British thermal unit 8
Bronchitis 182
Brown coal 18
Burner
  for oil 30
  for pulverised fuel 77

Calorie 8
Calorific value 8
  gross 10

Calorific value
  net 10
  of some fuels 11
Calorimeter 8, 10
Cancer of the lung 183
Cannel 19
Carbon 7, 11
Carbon bisulphide 118
Carbon dioxide 69, 112
  in flue gases 69
  in smog 175, 177
Carbonization 35, 84
Carbon monoxide 112
  from engines 49
  in blast furnace gases 87
Cascade impactor 143
Cellulose 7
Cement kiln 92
Central heating 97, 100
  Roman 96
Chain-grate stoker 73
Charcoal 15
Chemical works 117
Chimney height 178, 204, 211
Chlorine
  in coal 114
  in smog 117
Chlorofluorocarbons 50
CHP 101
Chromatography 146
Cigarette smoking 183
Clay industries 90
Clean Air
  Act 208, 209
  Zones, 97, 213
Coal
  analysis 16
  ash from 19
  bituminous 17, 18
  economy 106
  equivalent 44
  gas 40
  hazards 21
  macerals 16
  mineral matter in 20
  production 12
  rank 16
  stored 21
  sulphur in 20
  tar 38
  used 49
  varieties 17
  washed 20
Cochran boiler 58
Coke 11, 35
  gas 36
  low-temperature 37
  metallurgical 35
  oven 83
  spheres in grit 141
Coking stoker 74
Combustion 44, 69
Comfort zones 102
Conversions 220
Cooling towers 50
Corner tube boilers 66
Cupola 88
Cyclone 93, 121, 198

Daily cycle
  of smoke 166
  of sulphur dioxide 166
Daylight
  measurement 146
Denmark 215
Deposited pollution 130
  changes 164
  distribution 153
  rapid surveys 133
Deposit gauge 130
District heating 101
Domestic
  cooking 104
  heating 96
  hot water 100
Downdraught 171
Downdraught kiln 91
Downwash 171
Draeger tubes 145
Dust
  hazard in industry 121

Economic boiler 60
Electric
  furnace 80, 85
  heaters 102
  lighting 53
Electricity
  demand 55
  generation of 52
  hydro- 52
  uses 53
Electrolysis 54
Electrostatic precipitator 93, 121, 199
Energy 6
  consumption 44

Fats 7
Filters
  fabric 198
  high efficiency 198
Flue gas
  scrubbers 121, 191

Index

Flue gas
  velocity 111
Fluidised beds 79
Fluorine 113
  from chemical works 118
  in smog 177
Fly ash 141
Fog 172
Fossil fuel 7
France 215
Fuel
  artificial 34
  calorific value 11
  consumption 69
  fossil 7
  in World 11, 12
  measurement 8
  origin 6
  smokeless 15
Furnaces 80

Gas
  blast furnace 41
  coal 30
  fires 102
  natural 11, 12, 24, 33
  producer 40
  sewage sludge 41
  town 42, 81
  water 41
  works 81
Gasification 39
Germany 215
Gravity settling chambers 198
Greenhouse effect 113
Grit
  examination 140
  from coal 111
  from furnaces 89
  from industries 94
  prevention 195
Ground level concentrations 127, 170, 205

Hand firing 72
Health and Safety Executive 208
Heat
  balance 69
    in smog 175
  engine 46
  loss from houses 102
  storage cooker 106
Horizontal retort 82
Hot-water boilers 64
Homefire 38
House of the Future 103
Hydrocarbons 25
  synthesis 38
Hydrochloric acid 114, 117
Hydroelectricity 52
Hydrogen 11
Hydrogenation 38
Hydrogen sulphide
  from chemical works 118
  from furnaces 89
  offensive trades 121

Incinerators 120
Industrial
  boilers 59
  furnaces 80
Instantaneous water heaters 104
Insulation 102
Insulators 189
Internal combustion engine 47
Inversion 170
Ireland 216
Iron and steel works 86
Italy 216

Kenya 15
Kiln 91

Lagging 103
Lancashire boiler 59
Law and administration 207
Lead in petrol 113, 183
Lead dioxide instrument 138
Lighting 53
Lignin 7
Lignite 18
Lime kiln 92
Liquefaction 39
Los Angeles 114
Luxembourg 216

Materials damage 189
Mechanical stokers 73
Metallurgical coke 35
Metals damage 189
Meteorology 169, 205
Methane 33
Microscopic examination 140
Mining hazards 21
Motor vehicles 218

Natural gas 11, 12, 24, 33
Netherlands 216
Nitrogen oxides 51, 113, 114, 118
North Sea 24, 25

Nuclear
  plants 116
  power 115

Odours 115, 117
Offensive trades 121
Oil 12
  burner 30
  exploration 24
  fuel
    ash 28
    grades 29
    sulphur 28
  gasifier 32
  nozzle 31
  refinery 85
  storage 30
  trap 24
Open fire 98
  convector 99
  smokeless 98
Open-hearth furnace 90
Ovoids 38
Ozone 50, 114

Paint particles 117
PAN 115
Particles 121
  respirable 144
Peak load 56
Petroleum 23
  composition 25
  cracking 27
  distillation 26
  refining 24
  separation 26
  sulphur in 28
Photochemical smog 114
Photoelectric
  daylight instrument 146
  smoke reader 126
Phurnacite 19, 38
Plumes 109, 121, 171, 178, 204
Pneumoconiosis 182
Pollution rose 140
Portable
  smoke filter 128
  sulphur dioxide instrument 137
Power distribution 54
Primary pollutant 114
Producer gas 40
Public Health Act 207
Pulverized fuel 76
Pyrites 11

Radiation
  solar 6, 110
Radiators 100
Radioactive
  isotopes 6
  pollution 115
Railway engines 195
Rain 117, 169
Reciprocating steam engine 45
Recording instruments
  smoke 128
  sulphur dioxide 138
Recuperators 82
Refinery flare 85
Regenerators 83
Reheating furnaces 90
Respirable particulates 144
Retort
  horizontal 82
  vertical 84
Ringelmann chart 124, 210
Roomheater 99

Secondary pollutant 114
Sewage sludge gas 41
Shale oil 23
Ships 210
Significance tests 162
Smog
  and mortality 185
  constituents 173
  disasters 184
  Los Angeles 114
  photochemical 114
Smoke 109
  adhesion 110
  and daylight 156
  as an index of efficiency 71
  balance in smog 176
  control areas 213
    appliance grant 213
  deposition 110
  distribution
    by district 149
    in a town 153
  emission 109, 160, 210
  examination
    by microscope 140
    by electron microscope 143
  filter 126
    portable 128
    self-changing 128
    weighable 129
  from downdraught kilns 91
  from spoilbanks 119
  from steel industry 94
  in country districts 149

Index

Smoke 109
  in flue gases 71
  in smog 173
  irregular variation 169
  measurement 124
  particles 110
  penetration 110
  prevention 194
  stain on filter paper 129
  upward diffusion 156
  weekly cycle 166
  yearly cycle 166
Solid fuels 98, 105
Soot blowing 79
Space heating 54
Spectroscopy 146
Spoilbanks 22, 119
Spontaneous ignition 21
Sprinkler stoker 74
SST 50
Statistical methods 162
Steam
  engine 45
  pressure gauge 67
Steel industry 90
Stokers 73
Storage water heater 105
Strontium-90 117
Sulphate
  deposited 132
  distribution 148
Sulphur
  balance in smog 176
  in coal 111, 202
  in fuel oil 28
  in petroleum 25
  removal from fuel 202
Sulphur dioxide 111
  distribution
    by district 149
    in a town 153
    in the United Kingdom 158, 159
  emission 109, 160
  from chemical works 118
  from clay industries 94
  from oil refineries 85
  from spoilbanks 119
  from tall chimneys 112, 205
  in country districts 112
  in smog 173
  irregular variations 169
  measurement 135
  natural removal 112
  prevention 201
  removal from flue gases 203
  smell 112
  upward diffusion 156
  volumetric estimation 136

  weekly cycle 166
  yearly cycle 166
Sulphuretted hydrogen (*see* Hydrogen sulphide)
Sunshine 192
Suspended matter (*see* Smoke)
  examination by microscope 142

Television 56
Temperature 169
Thermal
  efficiency 69
  insulation 102
  precipitator 143
  storage
    boiler 61
    heating 102
Threshold limit value 113, 184, 209
Tidal power 53
Turbine
  gas 47
  steam 46
Turbulence 170
Tuyères 87

Ultraviolet radiation 50
Underfeed stoker 74
Uranium 12, 45
U.S.A. 219

Vegetation 188
Ventilating hoods 122
Vertical
  boiler 58
  retort 84
Viscosity 28
Visibility 191
Volatile matter 17

Wall
  cavity 103
  insulation 103
Washed coal 20
Water
  balance in smog 175
  gas 41
  heater
    domestic 104
    industrial 64
  tube boiler 57, 62
  vapour 50
Weekly cycle 166
West Germany 215
Wet scrubber 121, 199

Wind
  direction 169
  velocity 170
Wobbe number 42
Wood 11, 14

smoke 15
Works of art 190

Yearly cycle 166

DATE DUE

10/28